MULTI-VALUED AND UNIVERSAL BINARY NEURONS

Multi-Valued and Universal Binary Neurons
Theory, Learning and Applications

by

Igor N. Aizenberg
Neural Networks Technologies Ltd., Israel

Naum N. Aizenberg
Neural Networks Technologies Ltd., Israel

and

Joos Vandewalle
Departement Elektrotechniek, ESAT/SISTA,
Katholieke Universiteit Leuven, Belgium

KLUWER ACADEMIC PUBLISHERS
BOSTON / DORDRECHT / LONDON

A C.I.P. Catalogue record for this book is available from the Library of Congress.

ISBN 978-1-4419-4978-3

Published by Kluwer Academic Publishers,
P.O. Box 17, 3300 AA Dordrecht, The Netherlands.

Sold and distributed in North, Central and South America
by Kluwer Academic Publishers,
101 Philip Drive, Norwell, MA 02061, U.S.A.

In all other countries, sold and distributed
by Kluwer Academic Publishers,
P.O. Box 322, 3300 AH Dordrecht, The Netherlands.

Printed on acid-free paper

Contents

CHAPTER 3. *P*-REALIZABLE BOOLEAN FUNCTIONS AND UNIVERSAL BINARY NEURONS

CHAPTER 4. LEARNING ALGORiTHMS

CHAPTER 5. CELLULAR NEURAL NETWORKS WITH UBN AND MVN

CHAPTER 6. OTHER APPLICATIONS OF MVN AND MVN-BASED NEURAL NETWORKS

CHAPTER 7. CONCLUSIONS, OPEN PROBLEMS, FURTHER WORK

REFERENCES

INDEX

Acknowledgements

Part of the research for this book has been carried out at the SISTA/COSIC division of the Electrical Engineering Department (ESAT) of the Katholieke Universiteit Leuven. It was sponsored by a grant from the Belgian State Prime Minister's Office for Science, Technology and Culture (DWTC) for a visit of the first author at the KULeuven. We acknowledge also the support of the Concerted Actions project MIPS and MEFISTO 666 of the Flemish Community, the FWO project G.0262.97 "Learning and Optimization: an Interdisciplinary Approach", the Interuniversitary Attraction Pole IUAP P4-02 initiated by the Belgian State Prime Minister's Office for Science, Technology and Culture (DWTC) and the ESPRIT project DICTAM..

We acknowledge also the support of the company "Neural Networks Technologies Ltd." (Israel).

Preface

Artificial neural networks have attracted considerable attention in electrical engineering and computer science. Their reemergence stimulated new achievements in mathematics also. Research in the field of neural networks is intensifying, and more and more scientific groups from the large and famous research centers, and from the small and young universities are involved in this field. Many monographs and a huge number of journal papers are devoted to new results on neural networks and their applications. Hence the reader may wonder whether he/she can find anything new in this book.

First of all this book is not an overview or a generalization of existing and well-known results. The topic of this book is rather the deep analysis of two new (at least in the western literature) types of neural elements: multi-valued neurons, and universal binary neurons. We consider their mathematical model, learning algorithms for them, different types of networks for solving applied problems of image processing (cellular neural networks based on the considered type of neurons), and pattern recognition (some original types of neural networks, also based on the considered type of neurons). In fact the reader can be absolutely confident that he/she has never read the same in other books, may be except for the survey in Chapter 1 only.

Multi-valued and universal binary neurons are non traditional neural elements with high functionality. Their main common property is an operation with complex-valued weights, and therefore they have a complex-valued internal arithmetic, and similar activation functions, which are functions of the argument of the weighted sum. The high functionality of the universal binary neuron stems from the fact that the input/output map is not only described by a threshold, but by an *arbitrary* Boolean function. Hence it is much more general than threshold functions. Practically always it is

5

possible to implement the input/output mapping described by some partial-defined multiple-valued function on a *single* multi-valued neuron.

These neurons have two interesting properties, namely learning algorithms that quickly converge and broad functionality. Hence one can effectively apply neural networks based on multi-valued and universal binary neurons for the solution of the different challenging applied problems. We mention here in particular pattern recognition and image recognition. When only locally connected in a grid as in cellular neural network they apply very well to image filtering, extraction of details, and edge detection. An activation function of multi-valued neuron is used for developing powerful nonlinear filters. The high functionality of the multi-valued neuron, and the simplicity of its learning process have been used successfully for time-series prediction and signal interpolation by extrapolating orthogonal spectra.

Chapters 2 and 3 of the book are devoted to the mathematical background of the considered type of neurons, Chapter 4 deals with the learning behavior of these neurons. Chapters 2, 3 and part of 4 are significantly based on the PhD and ScD theses of the first and second author respectively. Both theses were written in Russian, and were not published in international journals. Hence this material is hardly known in the Western international scientific literature. The remaining chapters deal with more recent research results. The second part of Chapter 4 is devoted to a learning algorithm with error-correction rule. Chapters 5-6 deal with the neural networks for image processing and pattern recognition. They are based on the very recent results mainly obtained during very productive collaboration of all the authors during last several years, especially during 1 year and 7 month visit of the first author to K.U.LEUVEN in 1996-1998.

How to read this book

The book is addressed to all people who are interested in the fascinating field of neural networks or in their applications (mainly to image processing and pattern recognition).

The book starts with an introduction in Chapter 1. The text contains a deep mathematical part (Chapters 2-4), where mathematical background of multi-valued and universal binary neurons is presented, and simultaneously it contains a detailed description of solution of the interesting applied information processing problems (image processing and pattern recognition in Chapters 5-6).

We see at least three classes of potential readers, and we would like to give several recommendations in order to make the reading more interesting and rewarding.

The first set of readers are those who are interested in neural networks in general. From our point of view a deep mathematical background of multi-valued and universal binary neurons (together with proofs of all the theorems) presented in Chapters 2-4 will be most interesting for them. But we still encourage them to read also the Chapters 5-6 devoted to applications.

The second class of readers are those who are more interested in applications of neural networks. Clearly the applications to image processing and recognition, time-series prediction in Chapters 5 and 6 will be more interesting for them. However we advise these readers to also read the Chapters 2-4 by omitting the proofs of the theorems.

The third group of readers are those for whom a mathematical background of the neurons and neural networks, and their applications are equally interesting. We invite these readers to read the whole book with the same level of attention.

Finally we would like to add that the mathematical part of the book (especially proofs of the theorems) from the reader only requires basic knowledge of the algebra. The features of group characters, which often will be used, are considered within Section 2.1 of the Chapter 1 for the reader who is not familiar with this branch of the group theory.

Chapter 1

Introduction

This chapter is an initial point of the book. A brief historical observation of neural networks, their basic architectures, types of neurons, learning algorithms will be given in Section 1.1. We motivate the approach taken in this book in Section 1.2 by considering neurons with complex-valued weights, and especially multi-valued and universal binary neurons. A Chapter by Chapter overview of the book is given in Section 1.3. Own contributions are listed in Section 1.4.

1.   NEURAL NETWORKS AND NEURONS: BRIEF OBSERVATION

Among the different definitions of neural networks (NN), which one can find in many recent books we select the following attractive definition given by [Alexander & Morton (1990)], and adapted by [Haykin (1994)] as the most successful.

Definition 1.1.1. [Alexander & Morton (1990); Haykin (1994)] A *neural network* is a massively parallel distributed processor that has a natural propensity for storing experimental knowledge and making it available for use. It means that: 1) Knowledge is acquired by the network through a learning process; 2). The strength of the interconnections between neurons is implemented by means of the synaptic weights used to store the knowledge.

The learning process is a procedure of the adapting the weights with a learning algorithm in order to capture the knowledge. Or more mathematically, the aim of the learning process is to map a given relation between inputs and output (outputs) of the network.

Although the first artificial neural networks (ANNs) have been introduced as analogs of the biological neural networks they are quite far from each other. Of course, a human brain may be consider d as a nonlinear,

9

fast, and parallel computer (or more exactly, information-processing system), but it seems that we are far away from the creation of an artificial brain on the base of ANNs. Nevertheless many of ANNs applications are oriented to solution of the problems, which are natural for the human brain (the most complicate biological neural network): pattern recognition and classification, time series prediction, associative memory, control, robotics, financial market, etc. [Hecht-Nielsen (1988), Haykin (1994), Hassoun (1995)]. Thus, ANNs have been developed independently of their direct similarity (unsimilarity) to biological NN. They can now be considered as a powerful branch of present science and technology.

Let us present a brief historical point by point evolution of the field of neural networks. The following key points (especially from the point of view of topic of this book) may be considered (part of them are given in [Suykens et al. (1996)]).

- 1943 - [McCulloch and Pitts (1943)] - the first nonlinear mathematical model of the formal neuron, but without real applications.
- 1949 - [Hebb (1949)] - the first learning rule. One can memorize an object by adapting weights.
- 1958 - [Rosenblatt (1958)] - conception of the perceptron as a machine, which can learn and classify patterns.
- 1960-1962 - [Widrow & Hoff] - invention of adalines - a new type of neuroprocessors with unique features in signal processing and control.
- 1963 - [Novikoff (1963)] - significant development of the learning theory, a proof of the theorem about convergence of the learning algorithm applied to solution of the pattern recognition problem using perceptron.
- 1960-s - intensive development of the threshold logic, initiated by previous results in perceptron theory. A deep learning of the features of threshold Boolean functions, as one of the most important objects considered in the theory of perceptrons and neural networks. The most complete summaries are given in [Dertouzos (1965)], [Muroga(1971)].
- 1969 [Minsky, Papert (1969)] - potential limit of the perceptrons as computing system has been shown.
- 1977 [Aizenberg N., Ivaskiv (1977)] - general principles of multiple-valued threshold logic over the field of the complex numbers, as a deep generalization of the Boolean threshold logic principles to multiple-valued case.[1]
- 1977 [Kohonen (1977)] - consideration of the associative memory, as a content-addressable memory, which is able to learn.

---

[1] The book has been published in Russian only, and unfortunately the results, which are quite important, were not available in international scientific community. It was a great problem and misfortune for the scientists of the former USSR - to be isolated from all the word. Only a few authors could receive a special permit to publish some outstanding results abroad. These limitations have been partially lifted only recently - in 1992-1993.

- 1982 [Hopfield (1982)] - shows by means of energy functions that neural networks are capable to solve a large number of problems of the traveling salesman type. Revival of the research in the field.
- 1982 [Kohonen (1984)] - describes the self-organizing maps.
- 1985 [Aizenberg I. (1985)] - consideration of the so-called P-realizable Boolean functions over the field of the complex numbers, residue class ring and finite field. Model of the new neural element, which can learn and solve XOR-problem without network[2]
- 1986 [Rumelhart et al. (1986)] - a deep and comprehensive consideration of learning with backpropagation.
- 1988 [Chua and Yang (1988)] - Cellular neural networks (CNN). Each neuron in the CNN is connected with the neurons from its nearest neighborhood only. Interesting applications to image processing.
- Present - more and more scientists are involved to research in the field of neural networks and their applications. The existence of several international scientific societies, many high quality journals, more and more conferences confirm a high progress and nice perspectives of the field.

We noted here events and publications, which are important in general, but which are closely connected with the material of this book. Another historical overview of the best known neural networks architectures may be presented by Table 1.1.1 considered in [Suykens et al. (1996)], and adapted here.

All the architectures have an important common property. A neuron is a nonlinear processing element. Therefore a neural network, which consists of the interconnection of a collection of neurons, is also a nonlinear system. The non-linearity is defined by the activation function of the neuron, which defines the type of the corresponding neural element.

Let us describe very briefly a neuron in general and its most popular activation functions. A most general model of a neuron is presented in Fig. 1.1.1. A neuron has n inputs with input signals $x_1, ..., x_n$ (or synapses, or connecting links), and some synaptic weights $w_i, i=1, ..., n$. An extra weight w_0 does not correspond to any input, and often it is called *bias*. A central point of the neural processing is the evaluation of weighting sum $z = w_0 + w_1 x_1 + ... + w_n x_n$ of the input signals and then the evaluation of the neuron activation function $\varphi(z)$ for the value z of the weighted sum. This value of the activation function is the output of the neuron. If input/output mapping is described by some function $f(x_1, ..., x_n)$, the following equality holds

$$\varphi(z) = \varphi(w_0 + w_1 x_1 + ... + w_n x_n) = f(x_1, ..., x_n).$$

---

[2] The paper has been also published only in Russian, and unfortunately the results, which are quite important, were not available for international scientific community because of the reasons noted above.

Table 1.1.1. The most popular ANN architectures (year of introduction and their inventors/discoverers.

Year	Type of neural architecture	Inventor (Discoverer)
1943	Formal neuron	McCulloch, Pitts
1957	Perceptron	Rosenblatt
1960	Madaline	Widrow
1969	Cerebellatron	Albus
1974	Backpropagation network	Werbos, Parker, Rosenblatt
1977	Brain state in a box	Anderson
1978	Neocognitron	Fukushima
1978	Adaptive Resonance Theory	Carpenter, Grossberg
1980	Self-organizing map	Kohonen
1982	Hopfield net	Hopfield
1985	Bidirectional Associative Memory	Kosko
1985	Boltzman machine	Hinton, Sejnowsky, Szu
1986	Counterpropagation	Hecht-Nielsen
1988	Cellular neural network	Chua, Yang

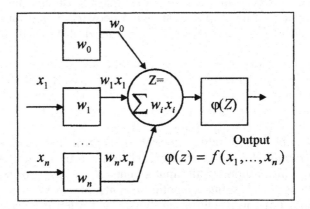

Figure 1.1.1. A general model of the neuron

A nonlinear activation function limits the amplitude of the output of a neuron. Usually the interval of the output and inputs of a neuron is [0, 1], or [1, -1]. We will consider in the following examples of the most popular types of activation function.

1. Threshold function

$$\varphi(z) = \text{sign}(z) = \begin{cases} 1, & \text{if } z \geq 0 \\ -1, & \text{if } z < 0 \end{cases}. \qquad (1.1.1)$$

2. Piecewise-linear function

$$\varphi(z) = \begin{cases} -1, & \text{if } z < -1 \\ z, & \text{if } -1 < z < 1 \\ 1, & \text{if } z > 1 \end{cases} \qquad (1.1.2)$$

3. Sigmoid function. Logistic function is example of the sigmoid function:

$$\varphi(z) = \frac{1}{1 + \exp(-az)}, \qquad (1.1.3)$$

where a is so-called slope parameter.

The definitions of activation functions are illustrated in Fig. 1.1.2.

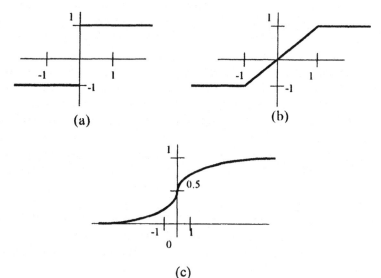

(a)

(b)

(c)

Activation functions: (a) - threshold, (b) - picewise, (c) - sigmoid

Figure 1.1.2.

There are some other types of the activation functions. For example, in applications of modeling and control hyperbolic tangent function $\varphi(z) = (1 - \exp(-z)) / (1 + \exp(-z))$ is normally used (it is an analog of sigmoid function, but it takes values in the interval [-1, 1] instead of [0, 1]), radial-basis functions (RBF) became recently popular as activation functions for MLP-neurons (see [Powell (1988)]).

Usually one distinguishes *feedforward* and *feedback* neural networks. A typical example of the feedforward network is multilayer perceptron (MLP) (see e.g., [Zurada (1992)], [Haykin (1994)]). MLP consists of an input layer of neurons, one or more hidden layers, and an output layer. The outputs of the neurons from a previous layer are connected with inputs of neurons of the following layer only (feedforward connections). In a feedback network the signal from the outputs of the neurons may transmit to the inputs of the neurons from the same layer (fully connected by such a way network is a *recurrent* network). The Hopfield network is an example. Another (and more important for us here) example of the neural network with feedback connections is the Cellular neural network ([Chua & Yang (1988)]. It is necessary to underline here the similarity with biological neural networks, where feedback connections are always presented (see e.g., [Freeman (1975)]).

To implement a mapping between inputs and outputs of a network, and on the other hand a mapping between inputs and output of each neuron (or in other words, according to the terms of Definition 1.1.1 to accumulate the knowledge) a learning process is used. Summarizing many definitions of learning beginning from [Hebb (1949)] and continuing by [Minsky (1961)], [Novikoff (1963)], [Natarajan (1991] and [Haykin (1994)] (many others may be added also) we can define learning in the following way. *Learning* is a process by which weighting parameters of a neural network are adapted through some process of the corrections to implement a corresponding mapping between inputs and output of the network (or between inputs and output of a neuron). The type of learning is determined by the method, in which the correction of weights is organized. In *supervised* learning a training (learning) set of input/output data is given, and the interconnection structure of the neurons is known in advance. Learning in such a case consists in iterative minimization of the error between the actual and desired outputs (with some given precision, or until precise coincidence of the actual and desired output vales) according to some learning rule for adapting the weights. So an "external teacher" is always present in supervised learning. In *unsupervised* learning desired outputs of the network are not available. In *reinforcement* learning the weights implementing a corresponding input/output mapping are obtained through a process of trials and errors. In context of this book we will only deal with the supervised learning.

It is evident from the equation (1.1.1) and Fig. 1.1.2a that the threshold activation function supposes inputs and outputs of the corresponding neuron and neural network are binary. This means that the input/output mapping for a neuron and a neural network in general are described by a Boolean function. Moreover, if we consider only one neuron, such a Boolean function has to be threshold, or linearly separable.

Definition 1.1.2. The Boolean function $f(x_1,...,x_n)$ is called a *threshold (linearly separable) function*, if it is possible to find such a real-valued weighting vector $W = (w_0, w_1,..., w_n)$ that equation

$$f(x_1,...x_n) = sign(w_0 + w_1x_1 + ... + w_nx_n) \qquad (1.1.4)$$

will be true for all the values of the variables x from the domain of the function f.

Of course, this is one of possible equivalent definitions, and it is possible to find many of them (see e.g., [Dertouzos (1965)], [Winder (1969)], [Muroga (1971)], etc.) Geometrical interpretation of the threshold functions is very simple: weights define an equation of a hyperplane, which separates "-1s" of the function from "1s" (or "0s" from "1s", if the classical Boolean alphabet {0, 1} is used). We have to make a remark. Here and further we will use the following correspondence between the Boolean alphabets {0, 1} and {1, -1}:

$$0 \rightarrow 1; \ 1 \rightarrow -1; \ y \in \{0, \ 1\}, \ x \in \{1, -1\} \Rightarrow x = 1 - 2y \qquad (1.1.5)$$

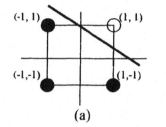 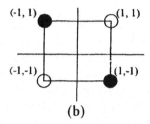

(a)   (b)

Figure 1.1.3. (a) - threshold function $f(x_1, x_2) = (1, -1, -1, -1)$,
"-1s" are separated from "1" by a line;
(b) - non-threshold function $f(x_1, x_2) = (1, -1, -1, \ 1)$ (XOR),
it is impossible separate "1s" from "-1s" by any line

Fig. 1.1.3a shows an example of threshold function (disjunction of two variables). Fig. 1.1.3b shows a typical and most popular example of non-threshold function (XOR, or *mod* 2 addition (or multiplication in alphabet {1, -1}) of two variables. Table 1.1.2 contains the values of variables and corresponding values of both functions.

Table 1.1.2 Threshold and not threshold functions

x_1	x_2	$\vee$	XOR
1	1	1	1
1	-1	-1	-1
-1	1	-1	-1
-1	-1	-1	1

The number of all Boolean functions of n variables is equal to 2^{2^n}, but the number of the threshold ones is substantially smaller. Really, for $n=2$ fourteen from sixteen functions (excepting XOR and *not* XOR) are threshold, for $n=3$ there are 104 threshold functions from 256, but for $n>3$ the following correspondence is true (T is a number of threshold functions of n variables):

$$\frac{T}{2^{2^n}} \underset{n>3}{\to} 0.$$

For example, for $n=4$ there are about 2000 threshold functions only from 65536 [Muroga (1971)]. To implement the non-threshold functions, different networks should be designed. One of the most popular types of network is MLP. We will not consider it in details, and anybody can find a deep consideration of MLP e.g., in [Zurada (1992)], [Haykin (1994)], etc.

So a very important observation, which we have to make, and which was one of the key starting points background point for the results presented in this book is the following: *The functionality of neural element with the threshold activation function is limited to threshold Boolean functions. To implement any non-threshold function a network should be designed.*

The natural questions, on which you will find a positive answer in this book, are the following: *Is it possible to design a neuron (or more exactly, an activation function), which could implement the non-threshold Boolean functions? Could such an implementation be similar to the implementation (1.1.4), i.e., is it possible to express a value of the function as a value of activation function evaluated on the value of the weighted sum?*

The activation functions (1.1.2) and (1.1.3) are oriented towards performing mappings not described by Boolean, but by band-limited functions. So inputs and outputs in such a case are multiple-valued in general. The main problem for the networks based on the neurons with such activation functions is organization of the learning. The original and one of the most popular learning rules for MLP in general is the backpropagation algorithm (see [Rumelhart et al. (1986)]). The backpropagation means that at each iteration of the learning process the actual outputs of all the neurons at all the layers should be evaluated and then compared with the desired outputs. Then the error, which is evaluated at the output layer is backpropagated from the output layer to input layer, and then weights could be adapted in such a way that error at the next iteration should be smaller than the previous one. So the backpropagation learning aims at minimizing the following cost function

$$\min_{w_{ij}^l} E = \frac{1}{P}\sum_{p=1}^{P} E_p, \quad E_p = \frac{1}{2}\sum_{j=1}^{N_L}(x_{j,p}^d - x_{j,p}^L)^2, \qquad (1.1.6)$$

where x_p^d is a desired output vector corresponding to the pth input pattern (or p^{th} learning subset), x_p^L is the corresponding actual output vector, L is a number of layers in MLP, N_l denotes a number of neurons at layer l, and E_p is the contribution of the pth pattern to the cost function E.

Problem (1.1.6) is a problem of nonlinear optimization because of nonlinearity of the activation function of the neuron. Many authors considered methods of its solution beginning from the pioneer paper [Rumelhart et al. (1986)]. It should be noted also e.g., [Narendra & Parthasarathy (1990), (1991)] (dynamic backpropagation for recurrent networks), [van der Smagt (1994)] (application of the conjugate gradient algorithms to solution of the learning problem). We will not deal with the details of nonlinear optimization here, because they are beyond the scope of this book (those who are interested may address e.g., to [Fletcher (1987)]). Learning of the RBF networks may be reduced to the application of the linear regression [Poggio & Girosi (1990)]. But the following remarks and questions are important. 1) Computer implementation of nonlinear optimization methods, also as linear regression methods is not a simple problem, and convergence of the corresponding algorithms sometimes is very slow. *2) Is it possible to consider the problem of implementation of the multiple-valued mappings from another point of view, namely, similar to implementation of Boolean mappings? 3) Is it possible in such a case to get significant increase of functionality of the single neuron, and thereby simplification of the learning algorithm?* Positive answers on both questions are given in this book.

After this brief relative positioning and formulation of the problems, we can move forward to clarify, how these problems may be solved.

2. PROBLEMS AND DIRECTIONS FOR THEIR SOLUTION

Problems, on which we will focus our attention, are the following: *1) Significant increase in functionality of the single neuron. 2) Elaboration of quickly converging learning algorithms with linear correction rules for the mappings described by non-threshold Boolean functions and functions of k-valued logic. 3) Application of the obtained results to the solution of applied problems in image processing and pattern recognition.*

After Minsky and Papert proved the limits of the single layer perceptrons [Minsky & Papert (1969)], further development of ANNs could to go in two directions. The first one includes different extensions of the perceptron architecture (e.g., MLP), or their generalization hoping to obtain the new results. The second one includes elaboration of more powerful neurons, first of all with new activation functions. The second direction may be connected with appearance of the Hopfield's paper [Hopfield (1982)], and further development considered in the previous Section. But in 1970th and later most authors worked in the first direction, first of all on VLSI implementations, solution of the applied problems using networks from perceptrons, and learning of the properties of threshold functions (see e.g. [Hill & Peterson(1974)], [Grossberg (1976)], [Kohonen (1978)]).

Unfortunately the international scientific community had a very limited access to the Russian publications at that time. Hence in the former Soviet Union science developed independently from the rest of the world. At the same time important results have been introduced.

In the paper [Aizenberg N. et al (1971a)] an idea of multiple-valued (k-valued) threshold logic, as generalization of Boolean threshold logic, has been proposed. This idea is based on the following considerations. If M is an arbitrary additive group, its power is not lower than k, and set $A_k = \{a_0, a_1, ..., a_{k-1}\}$, $A_k \subset M$ will be used as a structural alphabet, we will define the function P, which will be a k-valued predicate on the set M with the values in set A_k. A function $f: A_k^n \to A_k$ of n variables, where A_k^n is the nth Cartesian power of the set A_k, is a function of k-valued logic. A function $f(x_1, ..., x_n)$ of n variables will be a multiple-valued threshold function over the group M and relatively the predicate P, if such an $(n+1)$-dimensional vector $(w_0, w_1, ..., w_n)$ over the group M exists that $f(x_1, ..., x_n) = P(w_0 + w_1 x_1 + ... + w_n x_n)$. This approach has already been developed in [Aizenberg N. et al. (1971b, 1973)], and has become a theory of the k-valued threshold logic in [Aizenberg N. & Ivaskiv (1977)]. Actually, it was a wonderful background for developing a new type of neurons with high functionality, and new discrete-time neural networks. But unfortunately this idea has not triggered much attention in the eighties. Its rebirth took place in 1992 [Aizenberg & Aizenberg (1992)], where associative memory for storing gray-scale images has been proposed as a generalization of the binary associative memory ([Tan & Vandewalle (1990)] based on threshold neurons.

In the paper [Aizenberg I. (1985)] a mathematical background of the neural element with "complete functionality" (later we decided to call it *universal binary neuron* (UBN)) has been proposed. It was shown for the first time that the XOR problem could be solved on the single neuron by considering the complex domain. Schematically the following approach has been proposed: the weights are complex, and activation function of the neuron should be a function of the weighted sum (similar to idea of the multiple-valued threshold logic proposed in [Aizenberg N. & Ivaskiv (1977)]). For example, the activation function

$$\varphi(z) = \begin{cases} 1, & \text{if } 0 \le \arg(z) < \pi/2 \text{ or } \pi \le \arg(z) < 3\pi/2 \\ -1, & \text{if } \pi/2 \le \arg(z) < \pi \text{ or } 3\pi/2 \le \arg(z) < 2\pi \end{cases}$$

ensures solution of the XOR-problem for two inputs on the single neuron. Clearly, it is easy to check that for example, weighting vector $(0, 1, i)$, where i is an imaginary unity solves XOR-problem (see Table 1.2.1).

But this publication also has not been observed at that time. Further development of the idea generated the theory of P-realizable Boolean functions [Aizenberg I. (1991)], [Aizenberg & Aizenberg (1991)] and universal binary neurons [Aizenberg & Aizenberg (1993)], [Aizenberg N. et. al. (1993)].

Table 1.2.1. Solution of the XOR problem on the single universal binary neuron

x_1	x_2	$z = w_0 + w_1 x_1 + w_2 x_2$	$\varphi(z)$	XOR
1	1	$1+i$	1	1
1	-1	$1-i$	-1	-1
-1	1	$-1+i$	-1	-1
-1	-1	$-1-i$	1	1

So both ideas of multi-valued neuron and universal binary neuron have received a second life in 90th. Fortunately this second life is much more intensive and productive, especially during last years, when close collaboration between all three authors has began and developed (we will return to our own latest and recent publications in Section 1.4). On the other hand other research groups showed their interest to developing of the same field. For example, problems of the associative memory based on multi-valued neurons have been considered recently in [Jankovski et al. (1996)], general problem of CNN based on complex-valued neurons has been considered in [Toth et al. (1996)].

It should be noted that some authors considered some partially similar approach, but differed from the point of view of final results. Practically simultaneously in [Leung & Haykin (1991)] and [Georgiou & Koutsougeras] the complex domain backpropagation algorithms have been considered. But these authors considered the problem only for feed-forward networks (for MLP, in particular), and complexity of nonlinear programming problem (see above, Section 1.1) was a preventing factor in comparison with the simplicity of the learning algorithms, which will be presented below (see Chapter 4). Another direction is the development of multiple-valued threshold logic. In fact, the idea of representing the multiple-valued functions similarly to Boolean threshold functions is very attractive. The following contributions should be noted: [Moraga (1979), (1989)], [Si &

Michel (1995)], [Tanaka et al. (1996)]. The main idea, developed in the first two papers and separately considered later in the latter two papers with some supplements, is the concept of the multiple-valued logic over the field of the real numbers. But learning, for example, for multi-level sigmoidal function considered in the last paper, as a method to obtain a multi-valued effect, is too complicated in comparison with learning for multi-valued neurons and networks.

3.   CHAPTER BY CHAPTER OVERVIEW

So, what will the reader find in this book? To help the reader we would like to propose the following scheme of Chapters, their interconnections and dependencies.

General considerations about artificial neural networks, and an introduction to the problems considered in the book are presented in the *Chapter 1*.

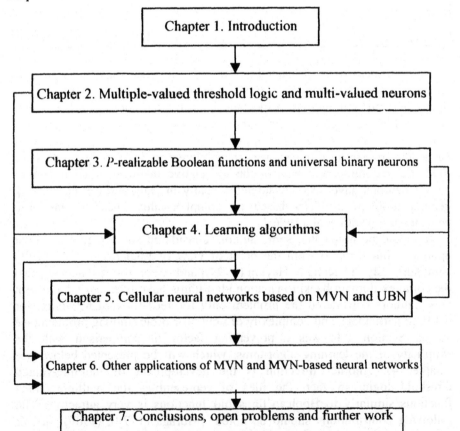

Chapter 2 is entitled *"Multiple-valued threshold logic and multi-valued neurons"*. The notion of multiple-valued (*k*-valued) function is introduced. Complex-valued weights and nonlinear activation function *P*, which is a function of the argument of the complex-valued weighted sum, are used for implementation of *k*-valued functions. Since properties of the group characters are used for investigation of the *k*-valued functions, they are considered in the beginning of the Chapter. It is shown that multiple-valued threshold logic over the field of the complex numbers considered here is a deep generalization of Boolean threshold logic for the case *k* >2. The properties of *k*-valued threshold functions are considered. Several necessary and sufficient conditions of the thresholding of *k*-valued function are proven. It is shown that the solution of the problem is partial function threshold, may be reduced to solution of the system of linear algebraic equations over the field of the complex numbers. Such a system may be obtained from the decomposition of the corresponding function of *n* variables by characters of the cyclic group of order k^n, which establish Chrestenson transformation. The notion of multi-valued neural element (MVN) as a neuron, which performs an input/output mapping described by a *k*-valued threshold function, is introduced. The topological and geometrical interpretation carried out by activation function *P* is investigated. A notion of *k*-edge in *n*-dimensional unitary space C^n is introduced. Synthesis of the multi-valued neuron, which is reduced to solution of the linear programming problem, is considered. Methods of synthesis for input/output mappings described by fully-defined and partially-defined *k*-valued functions are presented.

Chapter 3. P-realizable Boolean functions and universal binary neurons. It is shown that the mechanical expansion of the weight domain to the field of the complex numbers does not increase the number of Boolean threshold functions. To increase this number we propose a new nonlinear activation function P_B, which is a function of the argument of the complex-valued weighted sum (similarly to the MVN activation function). Also we introduce the notion of a *P*-realizable Boolean function of *n* variables as a function implemented by means of *n*+1 complex-valued weights and activation function P_B. On the base of this mathematical model the universal binary neuron (UBN) is introduced, as a neuron, which performs an input/output mapping described by *P*-realizable Boolean function. It is shown that the most popular non linear-separable XOR problem may be easily solved on a single UBN. Properties of *P*-realizable Boolean functions are investigated. It shown that many of them are generalization of the corresponding properties of Boolean threshold functions. In particular, generalized Chow theorems are proven, and the connection between weights of Boolean functions of *n*, *n*-1, *n*-2, ... 2, 1 variables is established. Different features of so-called *P*-odd and *P*-even activation functions are presented. Necessary and sufficient conditions of *P*-realizability for arbitrary Boolean function are proven. It is shown that the solution of the problem is a partial Boolean function *P*-realizable, may be reduced to the solution of a system of linear algebraic equations over the field of the complex numbers. Such a system may be obtained from the decomposition of the corresponding function of n variables by characters of the dyadic group of order 2^n, which establishes Walsh-Hadamard transform. *P*-realizable Boolean functions and mathematical model of UBN over residue class ring and finite field are considered also. Methods of synthesis of such a UBN are considered, and conjunctive (Reed-Muller) transformation is used for solution of the synthesis problem.

Chapter 4. Learning algorithms. This Chapter is completely devoted to learning algorithms for multi-valued and universal binary neurons introduced in Chapters 2 and 3. It is shown that learning of MVN is reduced to clarifying the problem of edge-separation of the learning subsets within the space C^n by a hyperplane, which equation is defined by the weighting coefficients (weights). The learning algorithm is considered step by step. Two linear learning rules - with correction coefficient, and with simple error correction are proposed. Theorems about convergence of the learning algorithm with both learning rules are proven. Optimal organization of the learning process is considered. Then learning algorithms for UBN are considered. It is shown that learning for UBN may be reduced to learning of MVN. On the other hand, learning algorithms especially for UBN, which are based on linear learning rules with correction coefficient and simple error-correction, are proposed. Theorems about convergence of these learning algorithms are proven. Finally, some problems of effective computer implementation of the learning algorithms without floating-point arithmetic, and for some special cases of activation functions are considered also.

Chapter 5. Cellular neural networks with UBN and MVN. A brief observation of CNN is included. It is shown that discrete-time CNN (DTCNN) have limited possibilities because they are based on the threshold neurons, and problems, which are described by non-threshold Boolean functions could not be solved using DTCNN. On the other hand the nonlinearity of the activation function of MVN may be successfully used for developing nonlinear filters, which may be used for noise reduction, and specific "anti-filtering" - amplification of medium and high frequency, which involves extraction even of the smallest details in image processing applications. These are the reasons for using UBN and MVN as basic CNN cells. Thus, CNN-UBN and CNN-MVN are considered, and effectively applied to the following image processing problems. The first of these problems is the problem of precise edge detection on binary and gray-scale images. Problems of edge detection corresponding to upward and downward brightness jumps, of the global edge detection, and edge detection by narrow directions are described by non-threshold Boolean functions. Using learning algorithms all the functions are implemented on the CNN-UBN. Simulation results, which confirm the high efficiency of the proposed method in comparison with classical algorithms, are presented. Along the same lines, i.e. through the description by non-threshold Boolean function, its learning, and CNN-UBN implementation, a problem of impulse noise filtering is solved. The concept of a multi-valued nonlinear filtering is proposed. It is shown that 2-D multi-valued filters may be realized using CNN-MVN. The templates for noise reduction with better performance than for order-statistic filters and for frequency correction are designed. The examples of noise reduction, correction of the blurred images, extraction of details on SAR and medical images are considered. A multi-valued filtering is proposed to be used for solution of the super resolution problem. It is very effective for final approximation of the spectral coefficients after an iterative procedure of the evaluation the coefficients corresponding to the highest frequency part of spectra. This allows the development of an algorithm and the evaluation on some examples.

Chapter 6. Other applications of MVN, and MVN-based neural networks. Different MVN-based neural networks are proposed for the solution of pattern recognition and associative memory problems. CNN-MVN associative memory for gray-scale images is considered. Then we show how so-called MVN-based neural network with random connections could be used as associative memory for gray-scale images. A simple single-layer network of MVN's is proposed for image recognition. The approach is based on learning and further analysis of the lowest part of orthogonal spectra. It is shown that such an approach is robust with respect to different changes and corruption of image (rotation, shifting, corruption by noise, blur, change of one image of object by another one, etc.). The example of face recognition is presented. It is shown that the high functionality of the single MVN may be used for time-series prediction. The example of currency exchange rate forecasting is considered. The same property of MVN is proposed to be used for solution of interpolation problem through extrapolation of orthogonal spectra to the highest frequency domain with prediction on MVN. Such an approach is compared with the approach based on multi-valued filtering approximation of spectra considered in Chapter 5.

The *Conclusions, open problems and further work* in the last chapter contains a brief overview of presented results and view to the open problems and future work in the direction developed in the book.

4. OWN CONTRIBUTIONS

The main part of the book (Chapters 2-6) contains results obtained by authors themselves only, little part of them in collaboration with some other colleagues (it is reflected in references list).

The following classes of results should be distinguished.

- Conceptual design of multiple-valued threshold logic over the complex numbers field based on the group-theoretical approach; properties of *k*-valued threshold functions; multi-valued neural element; synthesis of multi-valued neural element reduced to solution of linear programming problem; topological and geometrical interpretation of k-valued threshold functions [(Aizenberg N. & Ivaskiv (1977)], [Aizenberg & Aizenberg (1992, 1993, 1997)], [Aizenberg N. et al. (1993, 1995, 1996b)].

• Conceptual design of P-realizable Boolean functions over the complex numbers field, residue class ring, finite field; universal binary neuron; properties of P-realizable Boolean functions; solution of the XOR problem on the single neuron; synthesis of the universal binary neuron; generalization of the properties of threshold Boolean functions on the case of P-realizable Boolean functions; investigation of the connection between P-realizable functions of given and lower number of variables [Aizenberg I. (1985, 1991, 1997a)], [Aizenberg & Aizenberg (1991, 1993, 1994ab, 1997)], [Aizenberg N. et al. (1993, 1996a)].

• Learning algorithms for multi-valued and universal binary neurons; learning rules with correction coefficient and simple error correction; proofs of convergence of the corresponding learning algorithms; strategy of learning; computer implementation of the learning [(Aizenberg N. & Ivaskiv (1977)], [Aizenberg & Aizenberg (1992, 1993, 1994ab, 1997], [Aizenberg N. et al. (1995, 1996ab)].

• Cellular neural networks with multi-valued and universal binary neurons (CNN-UBN, CNN-MVN); solution of the precise edge detection problem and edge detection by narrow directions and their CNN-UBN implementation; impulsive noise filtering and its CNN-UBN implementation; multi-valued filtering, its CNN-MVN implementation and application to noise removal, frequency correction (extraction of details) and solution of the super resolution problem; design of a nonlinear cellular neural filtering algorithms [Aizenberg & Aizenberg (1992, 1993, 1994b, 1997, 1998)], [Aizenberg N. et al. 1995b, 1996a)], [Aizenberg I. (1997ab, 1998)], [Aizenberg I. & Vandewalle (1997) (1998)], [Aizenberg & Aizenberg & Vandewalle (1998abc)], [Aizenberg I. et. al. (1999)].

• Application of the MVN-based networks to the solution of the pattern recognition and associative memory problems; CNN-MVN as associative memory for gray-scale images; MVN-based network with random connections as associative memory; single-layer MVN-based network for the solution of pattern recognition problems based on analysis of orthogonal spectra, time-series prediction using single MVN; solution of interpolation problem through orthogonal spectra extrapolation on the single MVN [Aizenberg & Aizenberg (1992, 1994a)], [Aizenberg N. et al. (1995ab, 1996b)], [Aizenberg I. & Vandewalle (1997) (1998)], [Aizenberg & Aizenberg (1998) (1999)].

Chapter 2

Multiple-Valued Threshold Logic and Multi-Valued Neurons

In this Chapter multi-valued neurons are considered. The theoretical background of multi-valued neurons is theory of multiple-valued threshold logic over the field of the complex numbers. It is a deep mathematical generalization of Boolean threshold logic. Section 2.1 is devoted to a general approach for multiple-valued threshold logic, and group's characters as its main mathematical instrument. Section 2.2 is devoted to multiple-valued threshold functions over the field of the complex numbers and their features. The notion of multi-valued neuron as neural element, which implements input/output mapping described by multiple-valued threshold function is given in Section 2.3. We also present geometrical and topological interpretation of such a mapping. Section 2.4 is devoted to the synthesis of multi-valued neuron by linear programming method.

1. CHARACTERS OF THE FINITE GROUPS AND MULTIPLE-VALUED LOGIC

Multi-valued neural elements (MVN), which are one of the main subjects of this book, are mathematically based on the concept of multiple-valued threshold logic. In order to understand clearly the concept of an MVN and to grasp a high potential of such a type of neural element, we will explain here the main notions connected with multiple-valued threshold logic, and its main object the multiple-valued threshold functions. The main theoretical ideas of multiple-valued threshold logic built over the field of the complex numbers have been presented for the first time in [Aizenberg N et al. (1971)], and then developed in [Aizenberg N & Ivaskiv (1977)]. We would like to summarize here the main topics of previous publications, and also add some new material.

A general approach to multiple-valued (k-valued) threshold logic, which we will develop here, is based on the following background. Let M be an arbitrary additive group, and its order is not lower than k. Let $A_k = \{a_0, a_1, ..., a_{k-1}\}$ is an arbitrary set of a power k, which will be used as a structural alphabet. In particular it may be $A_k \subset M$.

Let us define a function P, which is a k-valued predicate on the set M with values within set A_k. A function $f: A_k^n \to A_k$ of n variables, where A_k^n is the nth Cartesian power of the set A_k, will be called a k-valued logic function. A function $f(x_1, ..., x_n)$ of n variables will be a multiple-valued threshold function over the group M and relative to the predicate P, if an $(n+1)$-dimensional vector $(w_0, w_1, ..., w_n)$ over group M exists such that

$$f(x_1, ..., x_n) = P(w_0 + w_1 x_1 + ... + w_n x_n).$$

Let set M be the additive group of the field of the real numbers R, A_2 is a two-elements subset $\{0, 1\}$ or $\{1, -1\}$ of the field R. Let also a function P be

defined   as $P(z) = \text{sign}(z) = \begin{cases} 1, & z \geq 0 \\ 0, & z < 0 \end{cases}$,   if $A_2 = \{0, 1\}$,   and

$\text{sign}(z) = \begin{cases} 1, & z \geq 0 \\ -1, & z < 0 \end{cases}$, if $A_2 = \{1, -1\}$. In this way the traditional notion of

Boolean threshold function is obtained (see Definition 1.1.2).

Consider now the additive group of the complex numbers field C as the group M. The letters of the k-valued structural alphabet A_k will be coded by k^{th} power roots of unity. It is well known that there are exactly k of the kth power roots of unity. The primitive k^{th} power root of unity is $\varepsilon = \exp(2\pi i/k)$, where i is an imaginary unity. So, the set of all the kth power roots of unity is $\widetilde{E}_k = \{\varepsilon^0 = 1, \varepsilon, ..., \varepsilon^{k-1}\}$. Exactly this set will be used as a set A_k. An univalent correspondence between the traditional k-valued structural alphabet $E_k = \{0, 1, ..., k-1\}$ of the *mod k* residues and the alphabet $\widetilde{E}_k$ is established by mapping $j \to \varepsilon^j$, $j=0, 1, ..., k-1$.

Thus, a function f realizing a mapping $f: \widetilde{E}_k^n \to \widetilde{E}_k$, where $\widetilde{E}_k^n$ is n^{th} Cartesian power of the set $\widetilde{E}_k$, is a function of k-valued logic (or multiple-valued function).

Evidently the sets $\widetilde{E}_k$ and E_k are the cyclic groups of the order k relatively to multiplication and *mod k* addition respectively. Many of the features of multple-valued threshold functions and multi-valued neurons, also as the features of P-realizable Boolean functions and universal binary neurons (Chapter 3) are based on the features of the finite group characters. Taking this fact into account, we will consider some features of the group characters in more details. The reader, who is familiar with these features, may skip the remainder part of this short Section.

Definition 2.1.1. A *character* of the finite Abelian group G is [van der Waerden (1971)] a homomorphism χ of the group G into multiplicative group C^* of the field of the complex numbers C. (one may also compare this definition with [Lang (1971)].

Let G be a group of an order $h = |G|$ defined as a direct product of the cyclic groups $G_1, G_2, ..., G_n$ of orders $h_1, h_2, ..., h_n$ respectively: $G = G_1 \times G_2 \times ... \times G_n$ (if the operation in the group G is additive then G is a direct sum of its subgroups: $G = G_1 \oplus G_2 \oplus ... \oplus G_n$). Each element of the group G may be presented uniquely: $g = g_1^{\xi_1} g_2^{\xi_{21}} ... g_n^{\xi_n}$, where g_j is a generating element of a cyclic group $G_j, j=1, 2, ..., n$, $0 \le \xi_j < h_j$, $h_j = |G_j|$ is the order of the group G_j. From the character's definition we obtain the following:

$$\chi(g_1^{\xi_1} g_2^{\xi_2} ... g_n^{\xi_n}) = (\chi(g_1))^{\xi_1} ... (\chi(g_n))^{\xi_n}. \qquad (2.1.1)$$

Evidently,

$$(\chi(g_j))^{h_{j_1}} = \chi(g_j^{h_j}) = \chi(e) = 1, \qquad (2.1.2)$$

where e is the unit element of the group G. It is clear that a value of the character χ on element g_j is some of the h_j-th roots of a unity. Let it be ε_j.

Thus, we can conclude from the definition of character and equations (2.1.1) and (2.1.2) that the value of the character χ on any element $g \in G$ can be defined in the following way:

$$\chi(g) = \chi(g_1^{\xi_1} g_2^{\xi_2} ... g_n^{\xi_n}) = \varepsilon_1^{\xi_1} \varepsilon_2^{\xi_2} ... \varepsilon_n^{\xi_n}. \qquad (2.1.3)$$

The inverse statement is also true. The equation (2.1.3) defines some character of a group G for any h_j-th root of a unity ε_j. Evidently the different combinations of the roots of a unity in (2.1.3) will establish the different characters. Since there are exactly h_j possibilities to choose ε_j in (2.1.3), the number of the different characters of a group G is equal to $h = h_1 h_2 ... h_n$. If all the ε_j in (2.1.3) are equal to 1, then we will obtain the main character χ_0 ($\chi_0(g) = 1$).

A product of characters is a product obtained by multiplication of two functions, e.g. $\chi_k \chi_l(g) = \chi_k(g) \chi_l(g)$. Evidently, a multiplication of two characters is a character, and for any character χ : $\chi\chi_0 = \chi_0\chi = \chi$. It is easy to check that the function $\chi^{-1}(g) = \chi(g^{-1})$ also is a character of a group G, and $\chi\chi^{-1} = \chi^{-1}\chi = \chi_0$. Therefore a set of the characters is Abelian group relatively to the character multiplication. This group is called a group of characters of the group G, and we will denote it $\chi(G)$.

If ε_j is a primitive h_j-th root of unity, then the character

$$\chi(g_1^{\xi_1} g_2^{\xi_2} \dots g_n^{\xi_n}) = \varepsilon_j^{\xi_1} \qquad (2.1.4)$$

is a generating element of the cyclic subgroup $\chi(G_j)$ of the group $\chi(G)$. Evidently, $|\chi(G_j)| = h_j$, and the group $\chi(G)$ is a direct product of the cyclic subgroups $\chi(G_j)$ of orders $h_1, h_2, \dots, h_n$ respectively: $c(G)=c(G_1)c(G_2) \dots c(G_n)$. It means that the group G and its group of characters $\chi(G)$ are isomorphic.

Let consider the following sum: $\sum_{g \in G} \chi(g) = \sum_{\xi_1=0}^{h_1-1} \dots \sum_{\xi_n=0}^{h_n-1} \chi(g_1^{\xi_1} g_2^{\xi_2} \dots g_n^{\xi_n})$.

Taking into account (2.1.3) we obtain the following:

$$\sum_{g \in G} \chi(g) = \sum_{\xi_1=0}^{h_1-1} \dots \sum_{\xi_n=0}^{h_n-1} \varepsilon_1^{\xi_1} \varepsilon_2^{\xi_2} \dots \varepsilon_n^{\xi_n} = \prod_{j=1}^{n} (1 + \varepsilon_j + \varepsilon_j^2 + \dots + \varepsilon_j^{h_j-1}) \quad (2.1.5)$$

But

$$1 + \varepsilon_j + \varepsilon_j^2 + \dots + \varepsilon_j^{h_j-1} = \begin{cases} h_j, & \text{if } \varepsilon_j = 1, \\ \dfrac{\varepsilon_j^{h_j} - 1}{\varepsilon_j - 1} = \dfrac{1-1}{\varepsilon_j - 1} = 0, & \text{if } \varepsilon_j \neq 1. \end{cases} \qquad (2.1.6)$$

If $\chi \neq \chi_0$, then at least one of $\varepsilon_j \neq 1$, and at least one of the multipliers in (2.1.5) is equal to zero according to (2.1.6). It means that $\sum_{g \in G} \chi(g) = 0$, if $\chi \neq \chi_0$. On the other hand, if $\chi = \chi_0$, then, taking into account the equalities (2.1.5) and (2.1.6), $\sum_{g \in G} \chi(g) = h$. Thus,

$$\sum_{g \in G} \chi(g) = \begin{cases} 0, & \text{if } \chi \neq \chi_0 \\ h, & \text{if } \chi = \chi_0 \end{cases}. \qquad (2.1.7)$$

The equation (2.1.7) involves the following:

$$\sum_{g, G} c_j(g) c_k(g^{-1}) = \begin{cases} 0, & \text{if } j \neq k, \\ h, & \text{if } j = k \end{cases}. \qquad (2.1.8)$$

Indeed, since $\chi_k(g^{-1}) = \chi_k^{-1}(g)$, and taking into account (2.1.7) we obtain:

$$\sum_{g \in G} \chi_j(g)\chi_k(g^{-1}) = \sum_{g \in G} \chi_j(g)\chi_k^{-1}(g) = \sum_{g \in G} = \chi_j \chi_k^{-1}(g) = \begin{cases} 0, & \text{if } \chi_j\chi_k^{-1} \neq \chi_0 \\ h, & \text{if } \chi_j\chi_k^{-1} = \chi_0, \end{cases}$$

and the equality (2.1.8) is true. This equality may be written in another form. Since $\chi_k(g^{-1}) = \chi_k^{-1}(g) = \overline{\chi}_k(g)$ (where $\overline{\chi}_k(g)$ is complex-conjugate with $\chi_{\kappa}(g)$), we obtain the following:

$$\sum_{g \in G} \chi_j(g)\,\overline{\chi}_k(g) = \begin{cases} 0, & \text{if } j \neq k, \\ h, & \text{if } j = k \end{cases} \tag{2.1.9}$$

The equality (2.1.9) is an orthogonality relation for the characters. It will be very important for us further.

Let G be an arbitrary set of a finite power $|G|=h$, and $F(G)$ is a set of all possible functions, which are defined on G, and take their values within field C. Evidently, a set $F(G)$ is a vecor space over the field of the complex numbers C with respect to the addition of the functions and multiplication of a function with a number. Let us transform a set $F(G)$ into unitary space. The scalar product of two functions within this space is defined in following way:

$$(f, \varphi) = \sum_{g \in G} f(g)\overline{\varphi}(g), \tag{2.1.10}$$

where $\overline{\varphi}(g)$ is complex-conjugated with $\varphi(g)$.

Theorem 2.1.1. A dimension of the unitary space $F(G)$ is equal to h.

Proof. Consider a set G with structure of Abelian group in some arbitrary way (it is possible, for example, to consider G as a cyclic group with a corresponding operation). Let $\chi_0, \chi_1, \ldots, \chi_{h-1}$ are all the characters of Abelian group G. According to the equalities (2.1.9) and (2.1.10) the characters $\chi_0, \chi_1, \ldots, \chi_{h-1}$ are pair-wise orthogonal vectors within the unitary space $F(G)$. Therefore the characters are linear independent over the field C. The theorem will be proven, if we will show that any function $f \in F(G)$ is a linear combination of all the characters $\chi_0, \chi_1, \ldots, \chi_{h-1}$. It should be noted that by such a way we will also prove that characters $\chi_0, \chi_1, \ldots, \chi_{h-1}$ form an orthogonal basis of the unitary space $F(G)$. Let us consider the following expression $\dfrac{1}{h}\sum_{j=0}^{h-1}(f, \chi_j)\chi_j(g_l)$, where (f, χ_j) is a scalar product, and $g_l \in G$ is fixed. Using the equality (2.1.10), on the base of the equality (2.1.9) we obtain the following:

$$\frac{1}{h}\sum_{j=0}^{h-1}(f,\chi_j)\chi_j(g_l)=\frac{1}{h}\sum_{j=0}^{h-1}\left(\sum_{g\in G}f(g)\,\overline{\chi_j}(g)\right)\chi_j(g_l)=$$

$$=\frac{1}{h}\sum_{g\in G}f(g)\sum_{j=0}^{h-1}\chi_j(g_l)\,\overline{\chi_j}(g)=\frac{1}{h}f(g_l)h\ .$$

It means that

$$f(g_l)=\frac{1}{h}\sum_{j=0}^{h-1}(f,\chi_j)\chi_j(g_l) \qquad (2.1.11)$$

for any $g_l \in G$, and theorem is proven.

Remark. It follows from the equalities (2.1.9) and (2.1.10) that the functions

$$\eta_0=\frac{1}{\sqrt{h}}\chi_0,\ \eta_1=\frac{1}{\sqrt{h}}\chi_1,\ ...,\ \eta_{h-1}=\frac{1}{\sqrt{h}}\chi_{h-1}$$

form an orthonormal basis of the unitary space. It is easy to conclude from the last fact and from the equality (2.1.11) that any function from $F(G)$ may be decomposed into Fourier series by functions $\eta_0,\eta_1,\ ...,\ \eta_{h-1}$:

$$f=\sum_{j=0}^{h-1}(f,\eta_j)\eta_j\ .$$

If the group G is a direct product of the cyclic groups of an order k then it is easy to conclude from the equality (2.1.3) that the value of any character of the group G on an arbitrary $g \in G$ is equal to some k^{th} power root of a unity, e.g. to some element of the set $\widetilde{E}_k$.

To understand many of the features of MVN and UBN, the sets $\widetilde{E}_k^n$ and E_k^n should be considered as multiplicative and additive Abelian groups respectively.

Let define a structure of multiplicative Abelian group on the set $\widetilde{E}_k^n$. A component by component multiplication of the vectors will be a group operation:

$$(\varepsilon^{\alpha_1},\varepsilon^{\alpha_2},...,\varepsilon^{\alpha_n})\circ(\varepsilon^{\beta_1},\varepsilon^{\beta_2},...,\varepsilon^{\beta_n})=$$

$$=(\varepsilon^{(\alpha_1+\beta_1)\bmod k},\varepsilon^{(\alpha_2+\beta_2)\bmod k},...,\varepsilon^{(\alpha_n+\beta_n)\bmod k}),$$

where ε is a primitive k^{th} power root of unity, and $\alpha_q,\beta_q=0,\ 1,\ ...,\ k-1;\ q=1,\ 2,\ ...,\ n$. The vectors, which components are k^{th} roots of unity always will be the results of such an operation. The vector $(\varepsilon^0,\ \varepsilon^0,\ ...,\ \varepsilon^0)=(1,\ 1,\ ....,\ 1)$ will be unit element of the group $(\widetilde{E}_k^n,\circ)$, which we will also denote simple $\widetilde{E}_k^n$, or $\widetilde{G}$.

Let us have a function $f(x_1, x_2, ..., x_n)$ of the k-valued logic of n variables, which implements the mapping $f: \widetilde{E}_k^n \rightarrow \widetilde{E}_k$. We can consider the variables $x_1, x_2, ..., x_n$ as functions defined on $\widetilde{G}$:

$$x_j = (\varepsilon^{\alpha_1}, ..., \varepsilon^{\alpha_j}, ..., \varepsilon^{\alpha_n}) = \varepsilon^{\alpha_j}, \quad j=1, 2, ..., n. \qquad (2.1.12)$$

According to (2.1.12) each of functions x_j becomes a character of the group $\widetilde{G}$. Really,

$$x_j(\varepsilon^{(\alpha_1+\beta_1)\bmod k}, ..., \varepsilon^{(\alpha_j+\beta_j)\bmod k}, ..., \varepsilon^{(\alpha_n+\beta_n)\bmod k}) = \varepsilon^{(\alpha_j+\beta_j)\bmod k} = \varepsilon^{\alpha_j\beta_j} =$$

$$= x_j(\varepsilon^{\alpha_1}, ..., \varepsilon^{\alpha_j}, ..., \varepsilon^{\alpha_n}) \times x_j(\varepsilon^{\beta_1}, ..., \varepsilon^{\beta_j}, ..., \varepsilon^{\beta_n}).$$

The vectors

$$g_1 = (\varepsilon, 1, 1, ..., 1),$$
$$g_2 = (1, \varepsilon, 1, ..., 1), ...,$$
$$g_n = (1, 1, ..., 1, \varepsilon)$$

are the generating elements of the cyclic subgroups for which the group $\widetilde{G}$ is a direct product. Taking into account the equality (2.1.4) we obtain the following: $x_j \chi(g_1^{\xi_1} g_2^{\xi_2} ... g_n^{\xi_n}) = \varepsilon_j^{\xi_1}$. It means that x_j is a generating element of the cyclic subgroup $\chi(G_j)$ of an order k of the group $\widetilde{G}$, and the group $\chi(G)$ is the following direct product: $\chi(G) = \chi(G_1) \times \chi(G_2) \times ... \times \chi(G_n)$. So by previous considerations the following theorem has been proven.

Theorem 2.1.2. The variables $x_1, x_2, ..., x_n$ of the function $f(x_1, x_2, ..., x_n)$ of k-valued logic considered as characters generate the group of characters of Abelian group $\widetilde{G}$.

Let G be an Abelian group, which is defined on the set $E_k^n = \{i_1, i_2, ..., i_n\}$, where $i_l \in E_k = \{0, 1, ..., k-1\}; l = 1, 2, ..., n$. The group operation within G is defined as *mod k* component by component addition. Let us describe the characters of groups G and $\widetilde{G}$.

Let consider lexicographic ordering within groups G and $\widetilde{G}$ in the following way. The number i, which is equal to

$$i = i_1 k^{n-1} + i_2 k^{n-2} + ... + i_{n-1} k + i_n, \quad i=0, 1, ..., k^{n-1} \qquad (2.1.13)$$

will be put to correspondence to each element $(i_1, i_2, ..., i_n) = g_i \in G$ (and to each element $(\varepsilon^{i_1}, \varepsilon^{i_2}, ..., \varepsilon^{i_n}) = \widetilde{g}_i \in \widetilde{G}$ respectively). The elements $g_i \in G$ and $\widetilde{g}_i \in \widetilde{G}$ will be ordered according to increasing of the numbers i.

An isomorphism between G and $\widetilde{G}$ will be defined by $\varphi : g_i \to \widetilde{g}_i$.

Let us create a canonical table of the characters for groups G and $\widetilde{G}$. The elements of groups will be presented as the columns according for increasing of numbers i. A value of each character χ on the any element $\widetilde{g} \in \widetilde{G}$ is evaluated according to the following equality:

$$\chi(\widetilde{g}) = \chi(g_1^{\xi_1} g_2^{\xi_2} ... g_n^{\xi_n}) = \varepsilon_1^{\xi_1} \varepsilon_2^{\xi_2} ... \varepsilon_n^{\xi_n}.$$

Since G and $\widetilde{G}$ are isomorphic $\chi(G) = \chi(\widetilde{G})$. Let each character $\chi_{j_1 j_2 ... j_n} = x_1^{j_1} x_2^{j_2} ... x_n^{j_n}$, which is considered as the k^n-dimensional vector-column, will be linked to index j in following way. Let $\chi_{j_1 j_2 ... j_n} = \chi_j$, where

$$j = j_1 k^{n-1} + j_2 k^{n-2} + ... + j_{n-1} k + j_n;$$

$$j_l = 0, 1, ..., k-1; \ l=1, 2, ..., n; \ j = 0, 1, ..., k^{n-1}. \tag{2.1.14}$$

By numbering the characters in this way the generating elements $x_1, x_2, ..., x_n$ of the character's group $\chi(\widetilde{G})$ will have the following numbers:

$$x_i = \chi_{k^n - i}. \tag{2.1.15}$$

We will obtain the canonical table of characters writing the characters χ_j for increasing numbers j (see the equalities (2.1.14)). To illustrate our approach the Table 2.1.1 contains a canonical table of characters for $k=3$ and $n=2$.

Let consider a canonical table of the characters as a squared matrix U_n of order k^n: $U_n = \|u_{ij}^{(n)}\|$, where $u_{ij}^{(n)} = \chi_j(g_i)$. For elements of matrix U_n the following equality is true:

$$u_{ij}^{(n)} = \chi_j(g_i) = x_1^{j_1} ... x_n^{j_n}(i_1, ..., i_n) =$$

$$= x_1^{j_1}(i_1, ..., i_n) ... x_n^{j_n}(i_1, ..., i_n) = \tag{2.1.16}$$

$$= \varepsilon_1^{i_1 j_1} ... \varepsilon_n^{i_n j_n} = \varepsilon^{i_1 j_1 + ... + i_n j_n}.$$

It is evident from the equality (2.1.16) that matrix U_n is symmetric.

It follows from the Theorem 2.1.1 that the dimension of the unitary space V of all the complex-valued functions defined on the group $\widetilde{G}$ is equal to k^n. The orthogonality relation (2.1.9) for characters may also be written as:

$$(\chi_i, \chi_j) = \sum_{g \in G} \chi_i(g)\overline{\chi}_j(g) = \begin{cases} 0, & \text{if } i \neq j \\ k^n, & \text{if } i = j \end{cases}. \tag{2.1.17}$$

Table 2.1.1. Canonical table of characters for $k=3$ and $n=2$.

G	$\widetilde{G}$	$\chi_0 =$ χ_{00}	$\chi_1 =$ χ_{01}	$\chi_2 =$ χ_{02}	$\chi_3 =$ χ_{10}	$\chi_4 =$ χ_{11}	$\chi_5 =$ χ_{12}	$\chi_6 =$ χ_{20}	$\chi_7 =$ χ_{21}	$\chi_8 =$ χ_{22}
$g_0 = (0,0)$	$\widetilde{g}_0 =$ $(\varepsilon^0, \varepsilon^0)$	1	1	1	1	1	1	1	1	1
$g_1 = (0,1)$	$\widetilde{g}_1 =$ $(\varepsilon^0, \varepsilon^1)$	1	ε	ε^2	1	ε	ε^2	1	ε	ε^2
$g_2 = (0,2)$	$\widetilde{g}_2 =$ $(\varepsilon^0, \varepsilon^2)$	1	ε^2	ε	1	ε^2	ε	1	ε^2	ε
$g_3 = (1,0)$	$\widetilde{g}_3 =$ $(\varepsilon^1, \varepsilon^0)$	1	1	1	ε	ε	ε	ε^2	ε^2	ε^2
$g_4 = (1,1)$	$\widetilde{g}_4 =$ $(\varepsilon^1, \varepsilon^1)$	1	ε	ε^2	ε	ε^2	1	ε^2	1	ε
$g_5 = (1,2)$	$\widetilde{g}_5 =$ $(\varepsilon^1, \varepsilon^2)$	1	ε^2	ε	ε	1	ε^2	ε^2	ε	1
$g_6 = (2,0)$	$\widetilde{g}_6 =$ $(\varepsilon^2, \varepsilon^0)$	1	1	1	ε^2	ε^2	ε^2	ε	ε	ε
$g_7 = (2,1)$	$\widetilde{g}_7 =$ $(\varepsilon^2, \varepsilon^1)$	1	ε	ε^2	ε^2	1	ε	ε	ε^2	1
$g_8 = (2,2)$	$\widetilde{g}_8 =$ $(\varepsilon^2, \varepsilon^2)$	1	ε^2	ε	ε^2	ε	1	ε	1	ε^2
			x_2		x_1					

The following theorem will be very useful for the consideration of the properties of multiple-valued threshold and P-realizable Boolean functions.

Theorem 2.1.3 The matrix $\dfrac{1}{\sqrt{k^n}} U_n = \widetilde{U}_n$ is symmetric and unitary, and for $k-2$ it is orthogonal.

Proof. According to the equality (2.1.16) the matrix U_n is symmetric, therefore the matrix $\widetilde{U}_n$ is symmetric also. To prove that it is unitary we have to show that $\widetilde{U}_n \widetilde{U}_n^* = I$, where I is an unity matrix, and matrix $\widetilde{U}_n^*$ is obtained from $\widetilde{U}_n$ by complex conjugation of all the elements. Taking into account the equality (2.1.16) we obtain the following:

$$\sum_{s=1}^{k^n} \frac{1}{\sqrt{k^n}} u_{is}^{(n)} \frac{1}{\sqrt{k^n}} \overline{u}_{is}^{(n)} = \frac{1}{k^n} \sum_{s=1}^{k^n} (\varepsilon_1^{i_1 j_1} .. \varepsilon_n^{i_n j_n}) \times (\varepsilon_1^{-i_1 j_1} ... \varepsilon_n^{-i_n j_n}), \quad s = (\gamma_1, ..., \gamma_n).$$

So $\dfrac{1}{k^n} \displaystyle\sum_{s=1}^{k^n} u_{is}^{(n)} \overline{u}_{is}^{(n)} = \dfrac{1}{k^n} \sum_{\gamma_1 ... \gamma_n} (\varepsilon_1^{i_1 \gamma_1} \varepsilon_1^{-j_1 \gamma_1}) ... (\varepsilon_n^{i_n \gamma_n} \varepsilon_n^{-j_n \gamma_n}) =$

$$= \frac{1}{k^n} \begin{cases} 1+1+...+1, & \text{if } (j_1,...,j_n) = (i_1,...,i_n), \text{e.g., if } i = j, \\ \sum_{\gamma_1,...,\gamma_n} \varepsilon_1^{(i_1-j_1)\gamma_1} \varepsilon_n^{(i_n-j_n)\gamma_n}, & \text{if } (j_1,...,j_n) = (i_1,...,i_n), \text{e.g., if } i \neq j, \end{cases} =$$

$$= \frac{1}{k^n} \begin{cases} k^n, & \text{if } i=j \\ \sum_{\gamma_1,...,\gamma_n} \chi_{(i_1-j_1,...,i_n-j_n)}(\gamma_1,...,\gamma_n), & \text{if } i \neq j \end{cases} = \begin{cases} 1, & \text{if } i=j, \\ 0, & \text{if } i \neq j \end{cases}$$

(because a sum over all the elements of the character's values is equal to zero for any character, which is different from main). It means that $\frac{1}{k^n} U_n U_n^* = I$, and therefore

$$U_n^{-1} = \frac{1}{k^n} U_n^*. \tag{2.1.18}$$

Hence the theorem is proven.

According to the orhogonality relation (2.1.17) the columns of the matrix U_n create the basis of the unitary space V, and columns of the matrix $\tilde{U}_n$ create the orthonormal basis of the same space.

The arbitrary function $f \in V$ may be decomposed into a Fourier series of the characters of the group G in following way:

$$f = \frac{1}{k^n} \sum_{i=0}^{k^n-1} (f, \chi_i) \chi_i. \tag{2.1.19}$$

The matrix U_n is Kronecker n^{th} power $U_n = U^{\otimes n}$ of the matrix

$$U = U_1 = \begin{pmatrix} 1 & 1 & . & 1 \\ 1 & \varepsilon & . & \varepsilon^{k-1} \\ . & . & . & . \\ 1 & \varepsilon^{k-1} & . & \varepsilon \end{pmatrix},$$

which represents a canonical table of the characters for a cyclic group of an order k. On the other hand the matrix U is the matrix of discrete Fourier transform of a dimension $k \times k$, and U_n is the matrix of Chrestenson transform [Chrestenson (1955)] of a dimension $k^n \times k^n$.

For $k=2$ we have: $U_1 = \begin{pmatrix} 1 & 1 \\ 1 & -1 \end{pmatrix}$, and since U_n in such a case contains Walsh functions [Walsh, (1923)], it will be a matrix of Walsh transform of a dimension $2^n \times 2^n$ in Hadamard ordering. The columns of matrix U_n in such a case will be all the characters of the group E_2^n (E_2^n is a group with respect to a component by component *mod 2* addition of *n*-dimensional binary vectors). For example, for $k=2$, $n=2$ we obtain the following (matrix of Walsh-Hadamard transform of a dimension 4 x 4):

$$\chi_{00} \ \chi_{01} \ \chi_{10} \ \chi_{11}$$

$$U_2 = \begin{pmatrix} 1 & 1 & 1 & 1 \\ 1 & -1 & 1 & -1 \\ 1 & 1 & -1 & -1 \\ 1 & -1 & -1 & 1 \end{pmatrix} \qquad (2.1.20)$$

$$x_2 \quad x_1$$

The generating elements of the characters group of the group E_2^n are the Rademacher functions [Rademacher (1922)]. The generating elements of the characters group of the group E_k^n will be called the generalized Rademacher functions.

Hereby the brief introduction to the group characters theory, and their connection with the subject of the book is completed.

2. MULTIPLE-VALUED THRESHOLD FUNCTIONS OVER THE FIELD OF THE COMPLEX NUMBERS

As it was mentioned above, our goal is to develop a neuron, which will perform all possible input/output maps described by multiple-valued (*k*-valued) functions.

Let $f(y_1,...,y_n)$ be a function of *n* variables in *k*-valued logic ($y_i \in \{0,1,...k-1\}$, $i=1, ..., n$). Let us consider the set of the kth roots of unity. We consider the univalent mapping $j \leftrightarrow \varepsilon^j$, where

$$\varepsilon^j = \exp(i2\pi j/k). \qquad (2.2.1)$$

According to the approach proposed above (see Section 2.1.1), we will code the values of *k*-valued logic by complex numbers, which are the kth roots of unity. In other words integer-valued variables $y_i \in \{0,1,...k-1\} = E_k$ are mapped onto complex-valued variables $x_i \in \{\varepsilon^0, \varepsilon, ..., \varepsilon^{k-1}\} = \tilde{E}_k$, $i=1, ..., n$.

Let *k* in (2.2.1) be fixed. Let us consider the following nonlinear function *P*, which is a function of the argument of the complex number *z* and is simultaneously a *k*-valued predicate defined on the field *C*:

$$P(z) = \exp(i2\pi j/k), \quad \text{if} \ \ 2\pi j/k \leq \arg(z) < 2\pi\,(j{+}1)/k. \quad (2.2.2)$$

In other words, if the complex plane is partitioned in k equal sectors by continuing of the rays corresponding to the kth roots of unity, and the argument of the number z is between arguments of the roots of unity ε^j and ε^{j+1}, then $P(z) = \varepsilon^j$. The formula (2.2.2) is illustrated in Fig. 2.2.1.

Definition 2.2.1. A function $f(x_1,...,x_n)$ of the k-valued logic is called a *multiple-valued threshold function*, if it is possible to find a complex-valued weighting vector $W = (w_0, w_1,..., w_n)$ such that equality

$$f(x_1,...x_n) = P(w_0 + w_1 x_1 + ... + w_n x_n) \quad (2.2.3)$$

(where P is the function (2.2.2)) holds for all values of the variables x_i from the domain of the function f.

We will also say in such a case that the weighting vector W *implements* the function f.

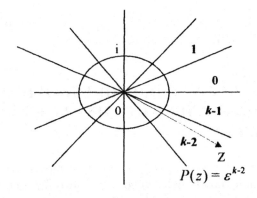

Figure 2.2.1. Definition of the activation function of the multi-valued
neuron (the equation (2.2.1))

Let us consider some examples of the threshold functions of k-valued logic (according to Definition 2.2.1). Let $k=3$, $n=2$, and our functions are defined by Tables 2.2.1 and 2.2.2, respectively.

Let us transfer a linear ordering on E_k $0<1<2< ... < (k-1)$ in a natural way onto $\tilde{E}_k$: $\varepsilon^0 < \varepsilon < ... < \varepsilon^{k-1}$ (such an ordering is obtained by arguments of the corresponding k^{th} roots of unity). The functions $f_1(z_1,z_2)$ and $f_2(z_1,z_2)$ are transformed to the well known Post functions $f_1(z_1,z_2) = \max(z_1,z_2)$; $f_2(z_1,z_2) = \min(z_1,z_2)$, but defined in alphabet $\tilde{E}_k$. It is easy to check that the weighting vector

$W = (-2 - 4\varepsilon,\ 4+5\varepsilon,\ 4+5\varepsilon)$ implements the function $f_1(z_1, z_2)$, and the weighting vector $W = (2+4\varepsilon,\ 5+4\varepsilon,\ 5+4\varepsilon)$ implements the function

Table 2.2.1.   **Table 2.2.2.**

z_1	z_2	$f_1(z_1,z_2)$
ε^0	ε^0	ε^0
ε^0	ε^1	ε^1
ε^0	ε^2	ε^2
ε^1	ε^0	ε^1
ε^1	ε^1	ε^1
ε^1	ε^2	ε^2
ε^2	ε^0	ε^2
ε^2	ε^1	ε^2
ε^2	ε^2	ε^2

z_1	z_2	$f_2(z_1,z_2)$
ε^0	ε^0	ε^0
ε^0	ε^1	ε^0
ε^0	ε^2	ε^0
ε^1	ε^0	ε^0
ε^1	ε^1	ε^1
ε^1	ε^2	ε^1
ε^2	ε^0	ε^0
ε^2	ε^1	ε^1
ε^2	ε^2	ε^2

$f_2(z_1, z_2)$.

It is possible to consider the following definition, which corresponds partially with the Definition 2.2.1. A function $f(x_1,\ldots,x_n)$ of the k-valued logic is said to be a threshold function, if it is possible to find a complex-valued weighting vector $W = (w_0, w_1, \ldots, w_n)$ such that

$$f(x_1, \ldots x_n) = w_0 + w_1 x_1 + \ldots + w_n x_n \quad \forall x \in dom\, f\,.$$

This means that argument of the weighted sum has to be equal to argument of some of the kth root of unity, which is a value of function f. Such a definition may be interesting because in the case $k=2$ we obtain one of the definitions of Boolean threshold function (one may compare with the Definition 1.1.2). On the other hand the number of functions that satisfy this condition is quite limited for $k>2$. Only the functions, which principally depend on single variable, are threshold for $k>2$. We will return to analysis of Definition 2.2.1 and the special case for $k=2$ below, when we will consider the Boolean case in detail (Chapter 3).

We will use below the following notions that are equivalent: $f(X) = f(x_1, \ldots, x_n)$,   where $X = (x_1, \ldots x_n)$, $W(X) = w_0 + w_1 x_1 + \ldots + w_n x_n$,   where $X = (x_1, \ldots x_n)$ and $W = (w_0, w_1, \ldots, w_n)$.

Let us consider some important features of the multiple-valued threshold functions, first of all some necessary and sufficient conditions for representation of the k-valued function described by (2.2.3).

Theorem 2.2.1. The function $f(X)$ of the k-valued logic is threshold with the weighting vector $W = (w_0, w_1, ..., w_n)$ if and only if a function $F(X): \tilde{E}_k^n \to C$ exists ($F(X) \neq 0$ for some X) such that

$$F(X)f(X) = W(X),$$ (2.2.4)

and

$$0 \leq \arg(F(X)) < \frac{2\pi}{k}$$ (2.2.5)

Proof. Necessity. Let $f(X)$ be a threshold function with the weighting vector W and α is an arbitrary vector from $\tilde{E}_k^n$. Let us define $F(X)$ in the following way. Let for all the $\alpha \in \tilde{E}_k^n$

$$|F(\alpha)| = |W(\alpha)|$$ (2.2.6)

and

$$\arg(F(\alpha)) = \arg(W(\alpha)) - \arg(f(\alpha))$$ (2.2.7)

It follows from the equality (2.2.7) that $\arg(W(\alpha)) = \arg(F(\alpha)) + \arg(f(\alpha))$, and from the last equality:

$$\arg(W(\alpha)) = \arg(F(\alpha)f(\alpha))$$ (2.2.8)

Since $|f(\alpha)| = 1$, and taking into account the equality (2.2.6), we obtain the following:

$$|W(\alpha)| = |F(\alpha)f(\alpha)|.$$ (2.2.9)

The following conclusion is involved from the equalities (2.2.8) and (2.2.9): $F(X)f(X) = W(X)$. This means that the equality (2.2.4) is true.

Let $f(\alpha) = \varepsilon^j$. Since $\dfrac{2\pi j}{k} \leq \arg(W(\alpha)) < \dfrac{2\pi(j+1)}{k}$ then

$$\frac{2\pi j}{k} - \arg(\varepsilon^j) \leq \arg(W(\alpha)) - \arg(f(\alpha)) < \frac{2\pi}{k},$$

and after simplification:

$$0 \leq \arg(W(\alpha)) - \arg(f(\alpha)) < \frac{2\pi}{k}.$$ (2.2.10)

Finally, taking into account the equalities (2.2.6) and (2.2.7), we complete the proof of necessity, because it is evident that the inequality (2.2.5) is also true.

Sufficiency. Let us have the function $f(X)$, the vector $W = (w_0, w_1, ..., w_n)$, the function $F(X)$, and the conditions (2.2.4)-(2.2.5) are true. We have to prove that $f(X)$ is a threshold function with the weighting vector W. Let the function $F(X)$ is built according the equalities (2.2.6)-(2.2.7). In such a case the equalities (2.2.8)-(2.2.9) are also true. It follows from the equality (2.2.7) that $\arg(F(X)) = \arg(W(X)) - \arg(f(X))$. Taking into account the last equation we can obtain from the inequality (2.2.5) the following:

$$0 \leq \arg(W(X)) - \arg(f(X)) < \frac{2\pi}{k}. \qquad (2.2.11)$$

Let $\alpha \in \widetilde{E}_k^n$ and $f(\alpha) = \varepsilon^j$. Since $\arg(\varepsilon^j) = \frac{2\pi j}{k}$, from the inequality (2.2.11) we obtain the following:

$$\frac{2\pi j}{k} - \arg(\varepsilon^j) \leq \arg(W(\alpha)) - \arg(f(\alpha)) < \frac{2\pi(j+1)}{k} - \arg(\varepsilon^j),$$ and from

the last inequality: $\frac{2\pi j}{k} \leq \arg W(\alpha) < \frac{2\pi(j+1)}{k}$ for any $\alpha \in \widetilde{E}_k^n$. This

means that $f(X)$ is a threshold function, and W is its weighting vector. The proof is completed.

Theorem 2.2.2. The function $f(X)$ of the k-valued logic is threshold with the weighting vector $W = (w_0, w_1, \ldots, w_n)$ if, and only if $W(X) \neq 0$ for any $X \in \widetilde{E}_k^n$, and such a function $\varphi(X): \widetilde{E}_k^n \to R$ exists, that

$$|W(X)|e^{i\varphi(X)} = W(X)\bar{f}(X), \qquad (2.2.12)$$

$$0 \leq \varphi(X) < \frac{2\pi}{k} \qquad (2.2.13)$$

($\bar{f}(X)$ is the complex-conjugate of $f(X)$).

Proof. Necessity. Let the function $f(X)$ of the k-valued logic be threshold with the weighting vector $W = (w_0, w_1, \ldots, w_n)$. According to the Theorem 2.2.1 a function $F(X): \widetilde{E}_k^n \to C$, ($F(X) \neq 0$ for any X) exists such that conditions (2.2.4) and (2.2.5) are true. Since $|f(X)| = 1$, it follows from the (2.2.4) that $|F(X)| = |W(X)|$. Let $\varphi(X) = \arg(F(X))$. Taking into account the inequality (2.2.5) we obtain $0 \leq \varphi(X) < \frac{2\pi}{k}$, and

$$F(X) = |F(X)|e^{i\varphi(X)} = |W(X)|e^{i\varphi(X)}.$$

Taking into account the last correspondence, equation (2.2.4) is transformed to the following:

$$|W(X)|e^{i\varphi(X)}f(X) = W(X). \qquad (2.2.14)$$

Let multiply both parts of the equation (2.2.14) by $\bar{f}(X)$. Taking into account that $f(X)\bar{f}(X)=1$, we obtain $|W(X)|e^{i\varphi(X)} = W(X)\bar{f}(X)$, and necessity is proven.

Sufficiency. Let the conditions of the theorem are true. Multiplying both parts of the equation (2.2.12) by $f(X)$ we will obtain the equation (2.2.14). Let $F(X)=|W(X)|e^{i\varphi(X)}$. Then

$$F(X)f(X) = W(X), \quad 0 \le \arg(F(X)) < \frac{2\pi}{k},$$

and function $f(X)$ is threshold, W is its weighting vector according to the Theorem 2.2.1

Lemma 2.2.1. Let $A = \{a_1, a_2, ..., a_m\}$ and $B = \{b_1, b_2, ..., b_m\}$ be finite sets, their elements are complex numbers, and for all the $j=1, 2, ..., m$ $|a_j|=|b_j|\neq 0$. If an interval $T = [-\tau, \tau]$ exists such that $\forall j = 1, 2, ..., m$ $\arg(a_j) \in T$ and $\arg(b_j) \notin T$, then $\sum\limits_{j=1}^{m} a_j \neq \sum\limits_{j=1}^{m} b_j$.

Proof. It is possible to put $\tau < \dfrac{\pi}{2}$, $-\pi \le \arg(z) < \pi$, $z \in C$ without loss of generality. Let the conditions of the lemma are true, but $\sum\limits_{j=1}^{m} a_j = \sum\limits_{j=1}^{m} b_j$.

Then $\operatorname{Re}\sum\limits_{j=1}^{m} a_j = \operatorname{Re}\sum\limits_{j=1}^{m} b_j$, and therefore:

$$\sum_{j=1}^{m} \operatorname{Re}(a_j) = \sum_{j=1}^{m} \operatorname{Re}(b_j), \qquad (2.2.15)$$

$$\sum_{j=1}^{m} |a_j|\cos(\arg(a_j)) = \sum_{j=1}^{m} |b_j|\cos(\arg(b_j)).$$

On the other hand $\forall j = 1, 2, ..., m$ $\cos(\arg(a_j)) > \cos(\arg(b_j))$; $|a_j|\cos(\arg(a_j)) > |b_j|\cos(\arg(b_j))$ and therefore

$$\sum_{j=1}^{m} |a_j|\cos(\arg(a_j)) > \sum_{j=1}^{m} |b_j|\cos(\arg(b_j)). \qquad (2.2.16)$$

The contradiction between (2.2.15) and (2.2.16) completes the proof of the lemma.

Theorem 2.2.3. The function $f(X)$ is a k-valued threshold logic function with weighting vector $W = (w_0, w_1, \ldots, w_n)$ if and only if a function $\varphi(X) : \widetilde{E}_k^n \to R$ exists such that

$$\left(\frac{W(X)}{|W(X)|} e^{-i\varphi(X)} \right)^k \equiv 1, \tag{2.2.17}$$

$$\sum_{\alpha \in \widetilde{E}_k^n} |W(\alpha)| e^{e^{i\varphi(\alpha)}} = \sum_{\alpha \in \widetilde{E}_k^n} |W(\alpha)| \bar{f}(\alpha), \tag{2.2.18}$$

$$0 \leq \varphi(X) < \frac{2\pi}{k}. \tag{2.2.19}$$

Proof. Necessity. Let $f(X)$ be a k-valued threshold logic function with the weighting vector $W = (w_0, w_1, \ldots, w_n)$. It follows from the Theorem 2.2.2 that such a function $\varphi(X) : \widetilde{E}_k^n \to R$ exists that equality (2.2.12) and inequality (2.2.13) are true. We obtain from (2.2.12) the following:

$$f(X) = \frac{W(X)}{|W(X)|} e^{-i\varphi(X)}.$$

Since $\left[f(X) \right]^k \equiv 1$, it means that the condition (2.2.17) is true. Adding both parts of the equality (2.2.12) with all $\alpha \in \widetilde{E}_k^n$ we obtain exactly the equality (2.2.18). The inequality (2.2.13) is true according to the Theorem 2.2.2, therefore the inequality (2.2.19) is also true.

Sufficiency. Let conditions of the theorem are true, but $W = (w_0, w_1, \ldots, w_n)$ is not the weighting vector for the function $f(X)$. It follows from the (2.2.18) that the function

$$f_1(X) = \frac{W(X)}{|W(X)|} e^{-i\varphi(X)} \tag{2.2.20}$$

defined on $\widetilde{E}_k^n$ is a k-valued logic function, and for any $\alpha \in \widetilde{E}_k^n$ $W(\alpha) \neq 0$. It is easy to obtain the following equality from the equality (2.2.20):

$$|W(X)| e^{i\varphi(X)} = W(X)\bar{f}_1(X). \tag{2.2.21}$$

A function $f_1(X)$ is threshold with the weighting vector W according to the Theorem 2.2.2, taking into account the equation (2.2.21). Simultaneously $f_1(X) \neq f(X)$ according to our assumption. Let $\Psi = \{ \alpha \in \widetilde{E}_k^n | \; |W(\alpha)| e^{i\varphi(X)} = W(\alpha)\bar{f}_1(\alpha) \}$, and $\Psi_1 = \widetilde{E}_k^n \setminus \Psi$. It is possible to prove that $\forall \; \alpha \in \Psi \; f(\alpha) = P(W(\alpha))$. It should be noted that $\Psi_1 \neq \varnothing$, because otherwise W will be a weighting vector for the function

$f(X)$ according to the Theorem 2.2.2. Let us transform the equation (2.2.18) to the following:

$$\sum_{\alpha \in \Psi} |W(\alpha)| e^{i\varphi(\alpha)} + \sum_{\alpha \in \Psi_1} |W(\alpha)| e^{i\varphi(\alpha)} = \sum_{\alpha \in \Psi} W(\alpha) \bar{f}(\alpha) + \sum_{\alpha \in \Psi_1} W(\alpha) \bar{f}(\alpha).$$

Since $\sum_{\alpha \in \Psi} |W(\alpha)| e^{i\varphi(\alpha)} = \sum_{\alpha \in \Psi} W(\alpha) \bar{f}(\alpha)$, then

$$\sum_{\alpha \in \Psi_1} |W(\alpha)| e^{i\varphi(\alpha)} = \sum_{\alpha \in \Psi_1} W(\alpha) \bar{f}(\alpha). \qquad (2.2.22)$$

The equality (2.2.21) involves the following:

$$\sum_{\alpha \in \Psi_1} |W(\alpha)| e^{i\varphi(\alpha)} = \sum_{\alpha \in \Psi_1} W(\alpha) \bar{f_1}(\alpha). \qquad (2.2.23)$$

From the equations (2.2.22) and (2.2.23) we obtain:

$$\sum_{\alpha \in \Psi_1} W(\alpha) f_1(\alpha) = \sum_{\alpha \in \Psi_1} W(\alpha) \bar{f_1}(\alpha). \qquad (2.2.24)$$

Let $\Psi_2 = \{\alpha \in \tilde{E}_k^n |\ f(\alpha) = f_1(\alpha),\ \alpha \in \Psi_1\}$, and $\Psi_3 = \{\alpha \in \tilde{E}_k^n |\ f(\alpha) \neq f_1(\alpha),\ \alpha \in \Psi_1\}$. It is evident that $\Psi_2 \cup \Psi_3 = \Psi_1$. It is impossible that $\Psi_3 = \varnothing$ because in such a case $\forall \alpha \in \Psi_1\ f(\alpha) = f_1(\alpha)$, and since $\forall\ \alpha \in \Psi\ f(\alpha) = P(W(\alpha)) = f_1(\alpha)$, thus $f(X) = f_1(X)$, which is impossible. Let us transform the equality (2.2.24) in the following way:

$$\sum_{\alpha \in \Psi_2} W(\alpha) \bar{f}(\alpha) + \sum_{\alpha \in \Psi_3} W(\alpha) \bar{f}(\alpha) = \sum_{\alpha \in \Psi_2} W(\alpha) \bar{f_1}(\alpha) + \sum_{\alpha \in \Psi_3} W(\alpha) \bar{f_1}(\alpha).$$

Taking into account that $\sum_{\alpha \in \Psi_2} W(\alpha) \bar{f}(\alpha) = \sum_{\alpha \in \Psi_2} W(\alpha) \bar{f_1}(\alpha)$, we obtain

$$\sum_{\alpha \in \Psi_3} W(\alpha) \bar{f}(\alpha) = \sum_{\alpha \in \Psi_3} W(\alpha) \bar{f_1}(\alpha). \qquad (2.2.25)$$

It follows from (2.2.21) that $\forall \alpha \in \Psi_3 \arg[W(\alpha) \bar{f_1}(\alpha)] = \varphi(\alpha)$. It means that

$$0 \leq \arg[W(\alpha) \bar{f_1}(\alpha)] < \frac{2\pi}{k}. \qquad (2.2.26)$$

Let suppose that for some $\alpha \in \Psi_3\ 0 \leq \arg[W(\alpha) \bar{f}(\alpha)] < \frac{2\pi}{k}$. Then taking into account that $|f(\alpha)| \equiv 1$, we obtain $W(\alpha) \bar{f}(\alpha) = |W(\alpha)| e^{i\omega(\alpha)}$, where $\omega(\alpha) = \arg[W(\alpha) f(\alpha)];\ 0 \leq \omega(\alpha) < \frac{2\pi}{k}$.   It   means   that

$f(\alpha) = P(W(\alpha))$, and therefore $f(\alpha) = f_1(\alpha)$, which contradicts with the condition $\alpha \in \Psi_3$. So $\forall \alpha \in \Psi_3$ and some $\tau < \dfrac{2\pi}{k}$

$$\arg\left[W(\alpha)\bar{f}_1(\alpha)\right] \in [0, \tau]; \quad \arg\left[W(\alpha)\bar{f}(\alpha)\right] \notin [0, \tau].$$

But additionally $\forall \alpha \in \Psi_3$ $|W(\alpha)\bar{f}_1(\alpha)| = |W(\alpha)\bar{f}(\alpha)| = |W(\alpha)| \neq 0$. According to the Lemma 2.2.1 $\forall k \in \{2,3,...\}$

$$\sum_{\alpha \in \Psi_3} W(\alpha)\bar{f}_1(\alpha) \neq \sum_{\alpha \in \Psi_3} W(\alpha)\bar{f}(\alpha).$$ We obtain a contradiction with the equality (2.2.25). It completes a proof of the theorem.

It should be noted that any complex-valued function $F(X)$ may be represented as follows:

$$F(X) = A(X) + B(X)\varepsilon , \qquad (2.2.27)$$

where $A(X)$ and $B(X)$ are some real-valued functions. In fsct, $\arg(\varepsilon) = \dfrac{2\pi}{k}$

because $\varepsilon = \exp(2\pi i / k)$. This means that the vector corresponding to the number ε on the complex plane is not collinear to the vector corresponding to the number $(1,0)$ which ensures the possibility of representation described by equality (2.2.27).

Theorem 2.2.4. The complex-valued function $F(X) = A(X) + B(X)\varepsilon \neq 0$, which is defined on $\widetilde{E}_k^n$ ($k>2$), satisfies the condition

$$0 \leq \arg(F(X)) < \frac{2\pi}{k} \qquad (2.2.28)$$

if and only if the real-valued functions $A(X)$ and $B(X)$ which are defined on $\widetilde{E}_k^n$ satisfy the conditions

$$A(X) > 0, \; B(X) \geq 0. \qquad (2.2.29)$$

The proof of this theorem is evident from the previous remarks concerning the possibility of the representation (2.2.27) for the function $F(X)$.

Corollary 2.2.1. The function $F(X)$ is threshold (for $k>2$) with the weighting vector $W = (w_0, w_1,..., w_n)$ if and only if such two real-valued functions $A(X)$ and $B(X)$, which are defined on $\widetilde{E}_k^n$ exist, that

$$| A(X) + \varepsilon B(X) | f(X) = W(X); \quad A(X)>0, B(X) \neq 0. \qquad (2.2.30)$$

A proof follows from the Theorems 2.2.1 and 2.2.4. The Corollary 2.2.1 is false for $k=2$ because the vectors corresponding to $\varepsilon = 1$ and $-\varepsilon = -1$ are collinear.

Let us consider some features of multiple-valued (k-valued) threshold functions, which are useful for classification of threshold functions. They are also interesting for checking if some function is a threshold function or not.

The features of the characters of group $\widetilde{E}_k^n$ considered in the Section 1.1 make it possible to establish the following important feature for k-valued threshold functions.

Let $F(X) = f(x_1,...,x_n)$ be a k-valued logic function of n variables. Let $(f, X_j) = b_j, j = 0,1, ..., n$, where X_j is a vector of values of jth variable (all the X_j are generalized Rademacher functions (see Section 2.1), and we will put $X_0 = \chi_0 = (1,1,...,1)$ for simplicity). Since the characters are orthogonal, and form a basis, a function f may be presented as:

$$f = \frac{1}{k^n}\sum_{j=0}^{k^n-1}a_j\chi_j = \frac{1}{k^n}(a_0 + a_1\chi_1 +...+a_{k^n-1}\chi_{k^n-1}).$$

Evidently, $b_0 = a_0$; $b_j = a_{k^{n-j-1}}$, $j = 1,...,n$. We will call the vector $b = (b_0, b_1, ..., b_n)$ the *characteristic vector* of function f.

It is interesting to mention that the characteristic vector of a multiple-valued threshold function (for any $k \geq 2$) contains the complete information about this function. In particular it contains information, which is sufficient for checking if it is a threshold function or not.

To prove this fact let us transform the right part of the equality (2.2.18) (see Theorem 2.2.3 above) in following way:

$$\sum_{\alpha\in\widetilde{E}_k^n}W(\alpha)f(\alpha) = (W(X),f(X)) =$$

$$= \frac{1}{k^n}\left((w_0 + w_1x_1 +...+ w_nx_n),(b_0 + b_1x_1 +...+ b_nx_n + \sum_{j\neq0,k^s;s=0,1,...,n-1}a_j\chi_j)\right).$$

Removing the parentheses under the scalar product in the right side of the last equality, we obtain the following:

$$\sum_{\alpha\in\widetilde{E}_k^n}W(\alpha)\bar{f}(\alpha) = w_0b_0 + w_1b_1 +...+ w_nb_n \ .$$

So taking also into account Theorem 2.2.3, we proved the following theorem:

Theorem 2.2.5. The function $f(X)$ is a k-valued logic threshold function with weighting vector $W = (w_0, w_1,..., w_n)$ if and only if a function $\varphi(X): \widetilde{E}_k^n \to R$ exists such that

$$\left(\frac{W(X)}{|W(X)|} e^{-i\varphi(X)} \right)^k \equiv 1 ,$$

$$\sum_{\alpha \in \tilde{E}_k^n} |W(\alpha)| e^{i\varphi(\alpha)} = w_0 b_0 + w_1 b_1 + \ldots + w_n b_n;$$

$$0 \le \varphi(X) < \frac{2\pi}{k} ,$$

where $b_0 = (f, \chi_0); \ b_j = (f, \chi_{k^{n-j-1}}), \ j = 1, \ldots, n$.

A generalization of two Chow theorems [Dertouzos (1965)] for the case of multiple-valued threshold functions then follows from the Theorem 2.2.5.

Theorem 2.2.6 (The first Chow Theorem). Let $f_1(x_1, \ldots, x_n)$ and $f_2(x_1, \ldots x_n)$ be functions of the k-valued logic. If the characteristic vectors of the functions f_1 and f_2 coincide then they both are threshold functions or they both are not threshold ones.

Proof. Let f_1 and f_2 be the functions satisfying conditions of the theorem, and $b = (b_0, b_1, \ldots, b_n)$ is their common characteristic vector. Suppose that the function f_1 is threshold with weighting vector W. Then a function $\varphi(x)$ exists such that it satisfies Theorem 2.2.5. Since $b = (b_0, b_1, \ldots, b_n)$ is a common characteristic vector for both functions f_1 and f_2 thus the conditions of the Theorem 2.2.5 are also true for the function f_2. It means that the function f_2 is threshold with the weighting vector W. If function f_1 is not threshold, the conditions of Theorem 2.2.5 are not true for it. But it means that they are also not true for the function f_2, and it is also not threshold. The theorem is proven.

It follows from the Theorem 2.2.3 that the weighting vector $W = (w_0, w_1, \ldots, w_n)$ is common for both functions f_1 and f_2. The second Chow theorem follows from this fact.

Theorem 2.2.6 (The second Chow Theorem). Let $f_1(x_1, \ldots x_n)$ and $f_2(x_1, \ldots x_n)$ be k-valued logic functions of n variables. If the functions f_1 and f_2 have a common characteristic vector, and one of them is threshold, then $f_1 = f_2$.

The following theorem is important for group classification of k-valued threshold functions.

Theorem 2.2.7. If the multiple-valued function (function of the k-valued logic) $f(x_1, \ldots x_n)$ is a threshold function (or not threshold), then the following functions are threshold (or not threshold) simultaneously:

$$f_1(x_1, \ldots, x_n) = f(x_1, \ldots, x_{i-1}, \sigma x_i, x_{i+1}, \ldots, x_n), \text{ where } \sigma^k = 1; \quad (I)$$

$$f_2(x_1,...,x_i,...,x_j,...,x_n) =$$
$$= f(x_1,...,x_{i-1},x_j,x_{i+1},...,x_{j-1},x_i,x_{j+1},...,x_n) ; \tag{II}$$

$$f_3(x_1,...,x_n) = \sigma f(\overline{\sigma}x_1,...,\overline{\sigma}x_n), \text{ where } \sigma^k = 1; \tag{III}$$

$$f_4(x_1,...,x_n) = x_j f(x_1\overline{x}_j,...,x_{j-1}\overline{x}_j,\overline{x}_j,x_{j+1}\overline{x}_j,...,x_n\overline{x}_j) . \tag{IV}$$

Before giving the proof it should be noted that all the possible transformations (I) of the set of k-valued functions of n variables form a group G_1 isomorphic to direct product of n copies of a cyclic group of an order k. All the possible transformations (II) form a group G_2 isomorphic to the symmetric group of a power n. All the possible transformations (III) form a group G_3 isomorphic to the cyclic group of order k. All the possible transformations (IV) form a group G_4 isomorphic to the direct product of the second order groups.

Proof of the Theorem 2.2.7.

1. Let $f(x_1,...x_n)$ be a threshold function. According to Theorem 2.2.4 a complex-valued function $F(X) = A(X) + \varepsilon B(X)$, which is defined on $\widetilde{E}_k^n$, exists such that

$$\big(F(X)f(X), \chi_l(X)\big) = 0, \text{if } \chi_l \neq X_i \ \forall i \in \{0,1,...,n\}; 0 \leq \arg(F(X)) < \frac{2\pi}{k}.$$

Let $\quad F_1(x_1,...,x_n) = F(x_1,...,x_{i-1},\sigma x_i,x_{i+1},...,x_n),\quad$ and $\alpha = (\alpha_1, ..., \alpha_n)$ pass through the group $\widetilde{E}_k^n$. Then $\alpha' = (\alpha_1, ...,\alpha_{i-1}, \sigma\alpha_i, \alpha_{i+1}, ..., \alpha_n)$ pass through the group $\widetilde{E}_k^n$ and

$$\big(F(x_1,...x_n)f(x_1,...x_n), \chi_l(x_1,...x_n)\big) =$$
$$= \big(F(x_1,...,x_{i-1},\sigma x_i,x_{i+1},...,x_n)f(x_1,...,x_{i-1},\sigma x_i,x_{i+1},...,x_n), \tag{2.2.31}$$
$$\chi_l(x_1,...,x_{i-1},\sigma x_i,x_{i+1},...,x_n)\big)$$

because the right and left sides of the equality (2.2.31) are differed by order of addends only. The following equality is true on the group $\widetilde{E}_k^n$:

$$(x_1,...,x_{i-1},\sigma x_i,x_{i+1},...,x_n) =$$
$$= (1, 1, ..., 1,\sigma,1, ..., 1)(x_1,...,x_{i-1},x_i,x_{i+1},...,x_n)$$

Therefore

$$\chi_l(x_1,...,x_{i-1},\sigma x_i,x_{i+1},...,x_n) = \chi_l(1,1,...,1,\sigma,1,...,1)\chi_l(x_1,...,x_i,...,x_n),$$

and the equality (2.2.31) is transformed to

$$\big(F(X)f(X), \chi_l(X)\big) =$$
$$= \overline{\chi}_l(1,1,...,1,\sigma,1,...,1) \times \big(F_1(X)f_1(X), \chi_l(X)\big) , \tag{2.2.32}$$

where $\overline{\chi}_l\,(1,\,1,\,...,\,1,\,\sigma,\,1,\,...,\,1) \neq 0$.

Let $\chi_i \neq X_i\,\forall i \in \{0,1,...,n\}$. Then $\left(F(X)f(X),\chi_l(X)\right) = 0$, and therefore $\left(F_1(X)f_1(X),\chi_l(X)\right) = 0$. Since the domain of the function $F(X)$ is the domain of the function $F_1(X)$ simultaneously, therefore $0 \leq \arg(F_1(X)) < \dfrac{2\pi}{k}$, and it means that the function f_1 is threshold according to the Theorem 2.2.4.

Let us consider the case, when function f is not threshold. Suppose that f_1 is a threshold function in such a case. But it means that the function $\tilde{f}_1(x_1,...x_n) = f_1(x_1,...,x_{i-1},\sigma^{-1}x_i,x_{i+1},...,x_n) = f(x_1,...x_n)$ is threshold. The last fact contradicts with the fact that the function f is not threshold. For the function (I) (case 1) theorem is proven.

Remark to the case 1. Let $f(X)$ is a threshold function, and the vector $W = (w_0, w_1,..., w_n)$ corresponds to the pair $(f,\ F)$, the vector $W' = (w'_0, w'_1,..., w'_n)$ corresponds to the pair (f_1, F). Then $w'_l = \dfrac{1}{k^n}\left(F_1(X)f_1(X), X_l\right)$, $l = 0, 1, ..., n$. According to the equality (2.2.32):

$$w'_l = \frac{1}{k^n}\,X_l\left(1,\,1,\,...,\,1,\,\sigma,\,1,\,...,\,1\right)\left(F(X)f(X), X_l(X)\right), \quad (2.2.33)$$

that is $w'_l = w_l$, if $l \neq i$, and $w'_i = \sigma w_i$, $i \in \{0, 1, 2, ..., n\}$. So

$$W' = (w_0,\ w_1,\ ...,\ w_{i-1},\ \sigma w_i,\ w_{i+1},\ ...,\ w_n).$$

Let us denote $b = (b_0, b_1,..., b_n)$, where $b_i = (f(X), X_i)$, $i = 0, 1, ..., n$.

By the same way: $b' = (b'_0, b'_1,..., b'_n)$, where $b'_i = (f_1(X), X_i)$, $i = 0, 1, ..., n$. It is easy to obtain from the equality (2.2.32) the following:

$$\left(f(X), X_l(X)\right) = \overline{X}_l\,(1,\,1,\,...,\,1,\,\sigma,\,1,\,...,\,1) \times \left(f_1(X), X_l(X)\right). (2.2.34)$$

The following correspondence between the characteristic vectors of functions f and f_1 is followed from the equality (2.2.34): $b'_l = b_l$, if $l \neq i$, and $b'_i = \sigma b_i$, e.g., $b' = (b_0,\ b_1,\ ...,\ b_{i-1},\ \sigma b_i,\ b_{i+1},\ ...,\ b_n)$.

2. Let the function $f(x_1,...,x_n)$ is threshold, and the vector $W = (w_0, w_1,..., w_n)$ corresponds to the pair $(f,\ F)$. Let $F_2(x_1,\ ...,\ x_i,\ ...,\ x_j,...,\ x_n) = F(x_1,\ ...,\ x_j,\ ...,\ x_i,...,\ x_n)$. Let $X_l(X) = x_1^{\gamma_1}x_2^{\gamma_2}...x_n^{\gamma_n}$, where $\gamma_i \in \{0,1,...k-1\}$ for all $i = \{1, 2, .., n\}$. Let

also $X_I(X) = x_1^{\gamma_1} x_2^{\gamma_2} \ldots x_i^{\gamma_i} \ldots x_j^{\gamma_j} \ldots x_n^{\gamma_n}$ be a character of the group $\widetilde{E}_k^n$,

and $X_I \neq x_t$ for any $t=0, 1, \ldots, n$, if $X_I \neq x_q$ for any $q=0, 1, \ldots, n$. Let

$\alpha = (\alpha_1, \ldots, \alpha_i, \ldots, \alpha_j, \ldots, \alpha_n)$ pass through the group $\widetilde{E}_k^n$. Then

$\alpha' = (\alpha_1, \ldots, \alpha_j, \ldots, \alpha_i, \ldots, \alpha_n)$ also pass through the group $\widetilde{E}_k^n$, and

$$\big(F(X)f(X), \chi_I(X)\big) = \Big(F_2(X)f_2(X), \chi_I(X)\Big)_{i,j} \qquad (2.2.35)$$

because the left and the right sides of the equality (2.2.35) are different only by order of addends within scalar products.

Let $\chi_I \neq x_q$ for any $q = 0, 1, \ldots, n$. Then

$$\Big(F_2(X)f_2(X), \chi_I(X)\Big)_{i,j} = \big(F(X)f(X), \chi_I(X)\big) = 0 \text{ because } \chi_I \neq x_t \text{ for}$$

any $t=0, 1, \ldots, n$. Evidently, $0 \leq \arg(F_1(X)) < \dfrac{2\pi}{k}$, and f_2 is a threshold

function according to the Theorem 2.2.4.

Let us consider a case, when $f(X)$ is not a threshold function. Let us suppose that $f_2(X)$ is threshold. According to the fact, which has been just proven, the function $\widetilde{f}_2(x_1, \ldots, x_i, \ldots, x_j, \ldots, x_n) = f(x_1, \ldots, x_n)$ has to be threshold, thus the contradiction is obtained, and the function $f_2(X)$ is not threshold. It completes proof for the function (II) (case 2).

Remark to the case 2. Let $f(X)$ be a threshold function, and the vector $W = (w_0, w_1, \ldots, w_n)$ corresponds to the pair (f, F), the vector $W' = (w_0', w_1', \ldots, w_n')$ corresponds to the pair (f_2, F). Then

$w_l' = \dfrac{1}{k^n}\big(F_2(X)f_2(X), X_l\big)$, $l = 0, 1, \ldots, n$. According to the equality

(2.2.35) $w_i' = w_j$; $w_j' = w_i$. Additionally $w_q' = w_q$ for any $q \notin \{i, j\}$. It is evident that $b_i' = b_j$; $b_j' = b_i$; $b_q' = b_q$, e.g.,

$b' = (b_0, b_1, \ldots, b_j, \ldots, b_i, \ldots, b_n)$;

$W' = (w_0, w_1, \ldots, w_j, \ldots, w_i, \ldots, w_n)$; $i, j \in \{1, 2, \ldots, n\}$.

3. Let $f(x_1, \ldots, x_n)$ is a threshold function, and the vector $W = (w_0, w_1, \ldots, w_n)$ corresponds to the pair (f, F). Let

$F_3(x_1,...,x_n) = F(\overline{\sigma}x_1,...,\overline{\sigma}x_n)$. If $\alpha = (\alpha_1, ..., \alpha_n)$ pass through the group $\widetilde{E}_k^n$, then $\alpha' = \overline{\sigma}\alpha$ also pass through the group $\widetilde{E}_k^n$, and

$$\sigma(F(X)f(X), \chi_I(X)) =$$
$$= \sigma(F(\sigma x_1,...,\sigma x_n)f(x_1,...,x_n), \chi_I(\sigma x_1,...,\sigma x_n)) \qquad (2.2.36)$$

Evidently, $(\overline{\sigma}x_1,...,\overline{\sigma}x_n) = (\overline{\sigma},...,\overline{\sigma})(x_1,...,x_n)$ within the group $\widetilde{E}_k^n$, so $\chi_I(\overline{\sigma}x_1,...,\overline{\sigma}x_n) = \chi_I(\overline{\sigma},...,\overline{\sigma})X_I(x_1,...,x_n)$. The equality (2.2.36) may be transformed to the following:

$$\sigma(F(X)f(X), \chi_I(X)) = \overline{X}_I(\overline{\sigma},...,\overline{\sigma})(F_3(X)f_3(X), \chi_I(X)), \qquad (2.2.37)$$

where $\overline{X}_I(\overline{\sigma},...,\overline{\sigma}) \neq 0$. Let $X_I \neq x_i$ for all $i=0$, 1, ..., n. Then $(F(X)f(X), \chi_I(X)) = 0$, and therefore $(F_3(X)f_3(X), \chi_I(X)) = 0$.

Evidently, $0 \leq \arg(F_1(X)) < \dfrac{2\pi}{k}$, and f_3 is a threshold function according to the Theorem 2.2.4.

Let us suppose that the function f is not threshold, and the function f_3 is threshold. But taking into account the last considerations, the function $\widetilde{f}_3(X) = \overline{\sigma}f_3(\sigma x_1,...,\sigma x_n) = f(X)$ has to be a threshold function. This fact contradicts with the condition that function f is not threshold.

Remark to the case 3. Let $f(X)$ be a threshold function, and the vector $W = (w_0, w_1,..., w_n)$ corresponds to the pair (f, F), the vector $W' = (w'_0, w'_1,..., w'_n)$ corresponds to the pair (f_3, F). Let now $\chi_I = X_i$, $i=0$, 1, ..., n in the formula (2.2.37). It is easy to obtain that $w'_0 = \dfrac{1}{k^n}(F_3(X)f_3(X), X_0) = \sigma w_0$, and $w'_i = w_i$ for all $i=1$, 2, ..., n. The following correspondence is following directly from the formula (2.2.37):

$$\sigma(f(X), \chi_I(X)) = \overline{\chi}_I(\overline{\sigma},...,\overline{\sigma})(f_3(X), \chi_I(X)). \qquad (2.2.38)$$

A global conclusion from the equality (2.2.38) is the following: $b' = (\sigma b_0, b_1, ..., b_n)$; $W' = (\sigma w_0, w_1, ..., w_n)$.

4. Let $f(x_1,...,x_n)$ be a threshold function, and the vector $W = (w_0, w_1,..., w_n)$ corresponds to the pair (f, F). Let $F_4(x_1, ..., x_n) = F(x_1\overline{x}_j, ..., x_{j-1}\overline{x}_j, \overline{x}_j, x_{j+1}\overline{x}_j, ...,x_n\overline{x}_j)$. Let $X_I(X) = x_1^{\gamma_1}x_2^{\gamma_2}...x_n^{\gamma_n}$, where $\gamma_i \in \{0,1,... k-1\}$ for all $i=\{1, 2, ..., n\}$. Let $\alpha = (\alpha_1, ..., \alpha_n)$ pass through the group $\widetilde{E}_k^n$. Then

$\alpha' = (\alpha_1 \overline{\alpha}_j, \ ...., \ \alpha_{j-1}\overline{\alpha}_j, \ \overline{\alpha}_j, \ \alpha_{j+1}\overline{\alpha}_j, \ ..., \ \alpha_n\overline{\alpha}_j)$ also pass through the group $\widetilde{E}_k^n$, and

$$\left(F(X)f(X), \chi_l(X)\right) = (\ F(x_1 x_j, ..., x_{j-1}x_j, x_j, x_{j+1}x_j, ..., x_n x_j) \times$$

$$\times f(x_1 x_j, ..., x_{j-1}x_j, x_j, x_{j+1}x_j, ..., x_n x_j), \tag{2.2.39}$$

$$\chi_l(x_1 x_j, ..., x_{j-1}x_j, x_j, x_{j+1}x_j, ..., x_n x_j))$$

because the left and the right sides of the equality (2.2.39) are different only by order of addends within scalar products. Taking into account that $f(x_1\overline{x}_j, \ ..., \ x_{j-1}\overline{x}_j, \ \overline{x}_j, \ x_{j+1}\overline{x}_j, \ ..., x_n\overline{x}_j) = \overline{x}_j f_4(x_1, ... x_n)$, the equality (2.2.39) may be transformed to the following:

$$\left(F(X)f(X), \chi_l(X)\right) =$$
$$= (\ F_4(X)\overline{x}_j f_4(X), \ \chi_l(x_1\overline{x}_j, ..., x_{j-1}\overline{x}_j, \overline{x}_j, x_{j+1}\overline{x}_j, ..., x_n\overline{x}_j)))'$$

or, in another form:

$$\left(F(X)f(X), \chi_l(X)\right) =$$
$$= (\ F_4(X)f_4(X), x_j \chi_l(x_1\overline{x}_j, ..., x_{j-1}\overline{x}_j, \overline{x}_j, x_{j+1}\overline{x}_j, ..., x_n\overline{x}_j))) \tag{2.2.40}$$

Let $\chi_l(X) = z_j \chi_l(x_1\overline{x}_j, ..., x_{j-1}\overline{x}_j, \overline{x}_j, x_{j+1}\overline{x}_j, ..., x_n\overline{x}_j) =$

$= x_1^{\gamma_1} ..., x_{j-1}^{\gamma_{j-1}} \ \overline{x}_j^{\gamma'_j} \ x_{j+1}^{\gamma_{j+1}} ..., x_n^{\gamma_n}$, where $\gamma'_j = 1 - \gamma_1 - \gamma_2 - ... - \gamma_n$. It is

clear that the mapping $\chi_l \overset{\varphi}{\to} \chi_l$ of the set of characters of the group $\widetilde{E}_k^n$ into itself is one-to-one, and $\varphi(x_0) = z_j$; $\varphi(x_q) = z_q$ for $q \notin \{0, j\}$, $\varphi^{-1} = \varphi$.

Let $X_l \neq x_l$ for all $i = 0, 1, ..., n$. Then, according to the equality (2.2.40) we obtain the following:

$$\left(F_4(X)f_4(X), \chi_l(X)\right) = \left(F(X)f(X), \chi_l(X)\right). \tag{2.2.41}$$

Evidently, $0 \leq \arg(F_1(X)) < \dfrac{2\pi}{k}$, and f_4 is a threshold function

according to the Theorem 2.2.4.

Let us suppose that f is not a threshold function, and that f_4 is a threshold function. But taking into account the fact, which has been just proven, the function

$\widetilde{f}_4(X) = x_j f_4(x_1\overline{x}_j, \ ..., \ x_{j-1}\overline{x}_j, \ \overline{x}_j, \ x_{j+1}\overline{x}_j, \ ..., x_n\overline{x}_j) = f(X)$ has to be

a threshold function. This fact contradicts with the condition that function f is not threshold.

Remark to the case 4. Let $f(X)$ be a threshold function, and the vector $W = (w_0, w_1, ..., w_n)$ corresponds to the pair $(f, \ F)$, the vector

$W' = (w'_0, w'_1, ..., w'_n)$ corresponds to the pair (f_4, F). Let now $\chi_i = X_i$, $i=0, 1, ..., n$ in formula (2.2.40). It is easy to show that

$$b' = (b_j, b_1, ..., b_{j-1}, b_0, b_{j+1}, ..., b_n);$$
$$W' = (w_j, w_1, ..., w_{j-1}, w_0, w_{j+1}, ..., w_n)$$

An important and very interesting class of multiple-valued functions is a class of the symmetric functions that is, of the functions, which are invariant relatively any permutation of the variables. For example, such classical functions as *max* and *min* in the alphabet $E_k = \{0, 1, ..., k-1\}$ are symmetric.

Let us consider function $f(x_1, x_2): M \times M = C^2 \to \widetilde{E}_k$ (M is an arbitrary set of the complex numbers).

Theorem 2.2.8. The symmetric function $f(x_1, x_2)$ is threshold if and only if the function of a single variable

$$F(x): M + M \to M \times M \ (M + M = \{x = x_1 + x_2 | x_1, x_2 \in M\}) \quad (2.2.42)$$

is threshold, and $F(x) = f(x_1, x_2)$, $x = x_1 + x_2$.

Proof. It should be noted first of all that $F(x'_1 + x'_2) = F(x_1 + x_2) = f(x_1, x_2)$, if $x'_1 + x'_2 = x_1 + x_2$.

Sufficiency. Let function $F(x)$, which satisfies the conditions of the theorem, exists and (w_0, w) is its weighting vector. The existence of such a function involves the following: $\forall (x_1, x_2) \in M \times M \ f(x_1, x_2) = F(x_1 + x_2) = F(x) = P(w_0 + wx) = P(w_0 + wx_1 + wx_2)$. It means that (w_0, w, w) is a weighting vector of function $f(x_1, x_2)$, and this function is threshold.

Necessity. Let $f(x_1, x_2)$ is a symmetric threshold function with the weighting vector (w_0, w_1, w_2). Then

$$f(x_1, x_2) = f(x_2, x_1) = P(w_0 + w_1 x_1 + w_2 x_2) = $$
$$= P(w_0 + w_2 x_1 + w_1 x_2) \quad (2.2.43)$$

It follows from the equalities (2.2.43) that the complex numbers $w_0 + w_1 x_1 + w_2 x_2$ and $w_0 + w_2 x_1 + w_1 x_2$ are not opposite each other as vectors at the complex plane, because they belong to the same semi-open sector of an angle value $2\pi / k$. It means that sum of such numbers is not equal to zero, and therefore

$$f(x_1, x_2) = f(x_2, x_1) = $$
$$= P\left(\frac{(w_0 + w_1 x_1 + w_2 x_2) + (w_0 + w_2 x_1 + w_1 x_2)}{2} \right). \quad (2.2.44)$$

Let $x = x_1 + x_2$, $w = (w_1 + w_2)/2$. It follows from the equality (2.2.44) that

$$f(x_1, x_2) = f(x_2, x_1) = P(w_0 + w_1 x_1 + w_2 x_2) = P(w_0 + wx) \overset{df}{=} F(x).$$

The last equation means that $F(x)$ is a threshold function, and (w_0, w) is its weighting vector.

The Theorem 2.2.8 may be easily generalized on the case of arbitrary number of variables: a symmetric function is threshold if and only if a corresponding function of one variable built like in (2.2.42) also is threshold.

3. MULTI-VALUED NEURAL ELEMENTS AND THEIR INTERPRETATION

Definition 2.3.1. [Aizenberg & Aizenberg (1992)] A *Multi-valued neural element* (*MVN*) is an element, for which the mapping from input to output is described by multiple-valued (*k*-valued) threshold function.

So MVN performs (2.2.3) taking into account (2.2.2) for a given mapping between input and output signals. Thus, MVN has n inputs and one output, and mapping f, which is implemented by MVN, is described by following formulas:

$$f(x_1, \ldots x_n) = P(w_0 + w_1 x_1 + \ldots + w_n x_n),$$

where $P(z) = \exp(i 2\pi j / k)$, if $2\pi j / k \leq \arg(z) < 2\pi (j+1)/k$.

It is clear that inputs and output of MVN are complex numbers, which are k^{th} roots of unity, or, which is the same, they are the elements of the set $\widetilde{E}_k^n$. Function P is the activation function of the MVN. A specific complex non-linearity of the function P has to be considered in more details in order to understand a potential of MVN.

In general, to understand the advantages of MVN for solution of the different applied problems it is necessary to clarify at least two important questions. One of them is synthesis of the neural element, or its learning. This problem will be considered in details in the Section 2.4 (synthesis), and in the Chapter 4 (learning). A problem of geometrical and topological interpretation of the MVN and k-valued threshold functions is reduced to investigation of the non-linearity of the function P. This problem will be considered now.

An interpretation of Boolean threshold functions is well-known. The weights of Boolean threshold function of n variables determine the coefficients of the hyperplane equation (within n-dimensional space over the field of the real numbers, see e.g., example in Section 1.1). A separation of the n-dimensional space over the field of the complex numbers is of course much more complicated problem. A precise understanding of its nature is very important for solving the learning problem, and for a successful solution of the different applied problems using multi-valued neurons.

Let us consider a class of functions of n complex variables, which includes the class of all the k-valued logic threshold functions.

Consider also the definition of the nonlinear function $P(z)$ (see (2.2.2)), which is not defined in the $z=(0,0)$. Later on in this Section we will select some precisely defined value from the set $\tilde{E}_k^n$ for $P(0)$.

It is possible to extend the Definition 2.2.1 of a k-valued threshold function in the following way. A complex-valued function $f(x_1,...,x_n)$, which is defined on the set $Z \subset C^n$, will be a k-valued threshold function, if it is a k-valued threshold function according to the Definition 2.2.1, and it is possible to find a complex-valued weighting vector $W = (w_0, w_1,..., w_n)$, and to define $P(0)$ in such a way that the equation

$$P\big(f(x_1,...x_n)\big) = P(w_0 + w_1 x_1 + ... + w_n x_n) \qquad (2.3.1)$$

(where P is the function (2.2.2)) will be true for all values of the variables x from the domain of the function f.

It is clear that if $(0, 0, ..., 0)$ is a weighting vector of a function f, then such a $w_0 \neq 0$ exists that $(w_0, 0, ..., 0)$ is also a weighting vector of a function f. Really, if $P(0) = \varepsilon'$, then $w_0 \neq 0$ should be chosen in such a way that $P(w_0) = \varepsilon'$. Moreover, the following statement is true.

Theorem 2.3.1. If domain $Z \subset C^n$ of the k-valued threshold function $f(x_1,...,x_n)$ is bounded, and $(w_0, 0, ..., 0)$ is its weighting vector then a complex number w_0' and a real number $\delta > 0$ exist such that for all the w_j, $j = 1,2,...,n$, for which $|w_j| < \delta$, the vector $(w_0', w_1, ..., w_n)$ also is a weighting vector of the function f.

Proof. Let $P(w_0) = \varepsilon'$. Then for all $X = (x_1,...,x_n) \in Z$: $P\big(f(X)\big) = P(w_0 + 0 \cdot x_1 + ... + 0 \cdot x_n) = \varepsilon'$, therefore $P\big(f(X)\big) = const$. Let $w_0' = \varepsilon^{t+\frac{1}{2}}$ which corresponds to the rotation of w_0 on the angle $\dfrac{\pi}{k}$. Since the domain of the function $f(x_1,...,x_n)$ is bounded, $N > 0$ exists such that $|x_j| < N, j = 1,2,,...,n$. Let $\delta = \dfrac{\sin(\pi / k)}{Nn}$, and taking all the $w_j, j = 1,2,...,n$ in such a way that $|w_j| < \delta$, we obtain the following:

$$\left| w_1 x_1 + \ldots + w_n x_n \right| \leq \left| w_1 \right| \| x_1 \| + \ldots + \left| w_n \right| \| x_n \| < N(\left| w_1 \right| + \ldots + \left| w_n \right|) < Nn\delta =$$

$$= \sin\left(\frac{\pi}{k}\right),$$

where $\sin\left(\dfrac{\pi}{k}\right)$ is the distance from the point w_0' until the bound of sector to

which it belongs. Therefore the sum of vectors (on the complex plane) corresponding to the complex numbers w_0' and $w_1 x_1 + \ldots + w_n x_n$ for any $(x_1, \ldots x_n) \in Z$ always will be within the same semi-open sector that the vector corresponding to w_0'. It means that $P(w_0') = P(w_0 + w_1 x_1 + \ldots + w_n x_n)$, and taking into account that $P(w_0') = P(w_0)$ it is easy to conclude that if $(w_0, 0, \ldots, 0)$ is a weighting vector then the vector $(w_0', w_1, \ldots, w_n)$ also is a weighting vector of the function f. Theorem is proven.

According to the Theorem 2.3.1, it is possible to find such a weighting vector $(w_0, w_1, \ldots, w_n)$ for a threshold function with a bounded domain that at least one of the components $w_1, \ldots, w_n$ is not equal to zero. It is also evident that that a weighting vector with the same property may be found not only for a k-valued threshold function with a bounded domain, but for function which has arbitrary domain with condition that $P(f(x_1, \ldots x_n)) \neq const$. Really, the last condition involves the following: it is possible to find such vectors $\alpha, \beta \in Z$, $\alpha = (\alpha_1, \ldots, c_n) \neq (\beta_1, \ldots, \beta_n) = \beta$ that $P(f(\alpha)) = P(W(\alpha)) \neq P(W(\beta)) = P(f(\beta))$. From the last expression it is evident that $W(\alpha) \neq W(\beta)$, and $W(\alpha) - W(\beta) = w_1(\alpha_1 - \beta_1) + \ldots + w_n(\alpha_n - \beta_n)$. Taking into account that at least one of the differences $\alpha_j - \beta_j \neq 0$, we have to conclude that at least one number from $w_1, \ldots, w_n$ is not equal to zero. The following theorem has been proven by last considerations.

Theorem 2.3.2. If at least one of the following conditions is true for the k-valued threshold function $f(x_1, \ldots, x_n)$ defined on the set $Z \subset C^n$: Z is bounded or $P(f(x_1, \ldots x_n)) \neq const$ then it is possible to find a weighting vector $W = (w_0, w_1, \ldots, w_n)$ for this function such that at least one of its the components $w_1, \ldots, w_n$ is not equal to zero.

So let one of the numbers $w_1, \ldots, w_n$ be nonzero. Let us consider a linear function $W(X) = w_0 + w_1 x_1 + \ldots + w_n x_n$; which is defined on

C^n without connection with some k-valued threshold function. Our goal is to clarify a geometrical interpretation of the set

$$\Pi_\varphi = \{\alpha \in C^n | \arg(W(\alpha)) = \varphi\}, \qquad (2.3.2)$$

where

$$\alpha = (\alpha_1,...,\alpha_n) \in C^n, \quad W(\alpha) = w_0 + w_1\alpha_1,...,w_n\alpha_n, \quad \varphi \in [0,2\pi[, \ w_j \neq 0$$

at least for one of $j=1, 2, ..., n$.

The equality $\arg(W(X)) = \varphi$ may be expressed as the equivalent system of equalities

$$\begin{aligned} \mathrm{Re}(W(X)) &= t\cos(\varphi) \\ \mathrm{Im}(W(X)) &= t\sin(\varphi) \end{aligned} \qquad (2.3.3)$$

where $t \in R$, and the equalities (2.3.3) are identical for $t > 0$.

Let $x_j = a_j + ib_j$, $j = 1,2,...,n$ and $w_j = u_j + iv_j$, $j = 0,1,...,n$, where i is the imaginary unit. Then the system (2.3.3) may be transformed to

$$\begin{aligned} u_0 + u_1a_1+...+u_na_n - v_1b_1-...-v_nb_n &= t\cos(\varphi) \\ v_0 + v_1a_1+...+v_na_n + u_1b_1+...+u_nb_n &= t\sin(\varphi) \end{aligned} \qquad (2.3.4)$$

After simple transformation of the system (2.3.4) (an elimination of the parameter t) we obtain the following linear equation:

$$\begin{aligned} &(u_1\sin\varphi - v_1\cos\varphi)a_1 +...+ (u_n\sin\varphi - v_n\cos\varphi)a_n - \\ &-(u_1\cos\varphi + v_1\sin\varphi)b_1 -...- (u_n\cos\varphi + v_n\sin\varphi)a_n + \\ &+ u_0\sin\varphi - v_0\cos\varphi = 0. \end{aligned} \qquad (2.3.5)$$

Taking into account that $w_j \neq 0$ for at least one of $j=1, 2, ..., n$, we can conclude that at least one of the coefficients under the variables $a_1,...,a_n,b_1,...,b_n$ is not equal to zero also. It means that rank of the system (2.3.5) is equal to 1.

Really, let us suppose that opposite is true, so that all the coefficients under the variables $a_1,...,a_n,b_1,...,b_n$ are equal to zero. Then we will obtain a system of $2n$ homogenous equations with $2n$ unknowns $u_1,...,u_n,v_1,...,v_n$ which will be decomposed into n pairs of the equations likes follows:

$$\begin{aligned} u_j\sin\varphi - v_j\cos\varphi &= 0, \\ u_j\cos\varphi + v_j\sin\varphi &= 0, \quad j = 1,2,...,n. \end{aligned} \qquad (2.3.6)$$

The determinant of the system (2.3.6) is equal to $\sin^2(\varphi) + \cos^2(\varphi) = 1$. Therefore $u_j = 0$, $v_j = 0$ for all the $j=1, 2, ..., n$. But it means that $w_1 = w_2 =...= w_n = 0$, which contradicts to choice of the w_j in (2.3.2).

Therefore we just proved that the equation (2.3.5) defines the hyperplane T_φ within the space R^{2n}. But the system (2.3.3) is not equivalent to the equation (2.3.5) for $t > 0$, and Π_φ does not coincide with the hyperplane T_φ.

Theorem 2.3.3. The plane $T^0 = \{Z \in C^n | W(X) = 0\}$ of a dimension $2n-2$ in R^{2n} separates the hyperplane T_φ into two semihyperplanes: the set Π_φ, which is the part of the hyperplane T_φ, where $t > 0$ and the part of T_φ, where $t < 0$. The plane T^0 is defined by the equations

$$\mathrm{Re}(W(X)) = 0,$$
$$\mathrm{Im}(W(X)) = 0, \tag{2.3.7}$$

or (which is equivalent):

$$u_0 + u_1 a_1 + \ldots + u_n a_n - v_1 b_1 - \ldots - v_n b_n = 0,$$
$$v_0 + v_1 a_1 + \ldots + v_n a_n + u_1 b_1 + \ldots + u_n b_n = 0. \tag{2.3.8}$$

Proof. First of all we have to show that T^0 is a plane within R^{2n}. It follows from the fact that the rank of the system of equations (2.3.8) is equal to 2. Really, since at least for single value of j $w_j \neq 0$, $j=1$, 2, ..., n, therefore the determinant

$$\begin{vmatrix} u_j & -v_j \\ v_j & u_j \end{vmatrix} = u_j^2 + v_j^2 = |w_j|^2 > 0$$

for some value of j. An inclusion $T^0 \subset T_\varphi$ is evident. Theorem is proven.

So the plane T^0 separates the hyperplane T_φ into two semi-hyperplanes: Π_φ (for $t > 0$ in (2.3.3)) and $T_\varphi | (\Pi_\varphi \cup T^0)$ (for $t < 0$ in (2.3.3)). It is also evident that T^0 does not depend on a value of φ (see (2.3.8)), and additionally $\forall \varphi \in [0, 2\pi[$ $T^0 \subset T_\varphi$. It means that

$$T^0 = \bigcap_{0 \le \varphi < 2\pi} T_\varphi. \tag{2.3.9}$$

It should be noted that T^0 is analytical plane [Schwartz (1967a)] of a dimension $n-1$ within space C^n because T^0 is defined by the linear equation $W(X) = 0$ in C^n, in other words the equation depends on variables $x_1, \ldots x_n$, and does not depend on the conjugate variables $\bar{x}_1, \ldots \bar{x}_n$.

Let T_1^0 be a linear subspace of a dimension $2n-2$ of the space R^{2n} corresponding to the plane T^0. In other words T_1^0 is parallel with T^0, contains the origin, and it is the space of solutions of the system, which consists of the two homogenous equations:

$$u_1a_1 + \ldots + u_na_n - v_1b_1 - \ldots - v_nb_n = 0,$$
$$v_1a_1 + \ldots + v_na_n + u_1b_1 + \ldots + u_nb_n = 0. \tag{2.3.10}$$

Let S^0 be the orthogonal complement [Schwartz (1967b)] of the space $T_1^{\prime 0}$, and the system of a $2n-2$ equations defines this complement:

$$s_1(a_1, \ldots, a_n, b_1, \ldots, b_n) = 0,$$
$$\ldots\ldots\ldots\ldots\ldots\ldots\ldots\ldots\ldots \tag{2.3.11}$$
$$s_{2n-2}(a_1, \ldots, a_n, b_1, \ldots, b_n) = 0.$$

Evidently, S^0 is a two-dimensional subspace, and therefore it is a plane (two-dimensional) within R^{2n}.

Theorem 2.3.4. The S^0 and T^0 have a single common point M^0.

Proof. Intersection of the planes S^0 and T^0 is defined by following system of linear equations:

$$u_0 + u_1a_1 + \ldots + u_na_n - v_1b_1 - \ldots - v_nb_n = 0,$$
$$v_0 + v_1a_1 + \ldots + v_na_n + u_1b_1 + \ldots + u_nb_n = 0,$$
$$s_1(a_1, \ldots, a_n, b_1, \ldots, b_n) = 0, \tag{2.3.12}$$
$$\ldots\ldots\ldots\ldots\ldots\ldots\ldots\ldots\ldots$$
$$s_{2n-2}(a_1, \ldots, a_n, b_1, \ldots, b_n) = 0$$

that is a result of a union of the systems of linear algebraic equations (2.3.8) and (2.3.11), which define the planes T^0 and S^0, respectively.

Since a rank of the matrix of the system (2.3.12) is equal to $2n$ (the sum of dimensions of the subspace $T_1^{\prime 0}$ and its orthogonal supplement S^0), and this system contains exactly $2n$ unknowns, it has single solution $M^0 = (a_1^0, \ldots, a_n^0, b_1^0, \ldots, b_n^0)$.

Theorem 2.3.5. The plane S^0 of a dimension 2 intersects with the semi-hyperplane $\overline{\Pi}_\varphi = \Pi_\varphi \bigcup T^{\prime 0}$ by ray I_φ starting from the point M^0.

Proof. Intersection of the plane S^0 with the hyperplane T_φ' is a plane within R^{2n}, which is defined by following system of linear algebraic equations:

$$s_1(a_1,...,a_n,b_1,...,b_n) = 0,$$

$$...............$$

$$s_{2n-2}(a_1,...,a_n,b_1,...,b_n) = 0,$$

$$(u_1\sin\varphi - v_1\cos\varphi)a_1 + ... + (u_n\sin\varphi - v_n\cos\varphi)a_n -$$

$$-(u_1\cos\varphi + v_1\sin\varphi)b_1 - ... - (u_n\cos\varphi + v_n\sin\varphi)a_n +$$

$$+u_0\sin\varphi - v_0\cos\varphi = 0$$

(2.3.13)

that is the result of the union of the systems of linear algebraic equations (2.3.11) and (2.3.5), which define the plane S^0 and the hyperplane T'_φ, respectively. The rank of the system (2.3.13) is equal to $2n$-1 for any $\varphi \in [0, 2\pi[$. It means that all the equations of this system are linearly independent. Really, suppose the opposite. Let the equations of (2.3.13) are linearly dependent. Since the equation (2.3.5) is a linear combination of the equations of (2.3.8), the linear dependence of the equations of (2.3.13) follows to the conclusion that the equations of the system (2.3.12) also are linearly dependent. The last conclusion contradicts with the fact, which has been shown in the proof of the Theorem 2.3.4. This means that the rank of the system (2.3.13) is equal to $2n$-1, and therefore this system defines a plane of a dimension 1, that is a line. Finally, the semi-hyperplane Π_φ is separated from the rest part of the hyperplane T_φ by plane T^0, which intersects with the plane S^0 in the single point M^0. Therefore a part of the line defined by (2.3.13) forms the ray $L_\varphi = S^0 \cap \Pi_\varphi$ starting from the point M^0. Theorem is proven.

Let us choose a polar coordinate system with the center M^0 and the polar axis L_0 $(\varphi = 0)$ within the two-dimensional plane S^0.

We will consider the smaller angle from the two angles between the rays L_φ and L_ψ as angle between these rays. Since k ≥ 2, therefore value of the angle between L_φ and L_ψ is always smaller than π.

Theorem 2.3.6. A value of the angle between the polar axis L_0 and the ray L_φ is equal to φ.

Remark. Evidently, the angle between the semi-hyperplanes Π_φ and Π_ψ is measured by linear angle between L_φ and L_ψ. So, this theorem is equivalent to the following one: the angle between the semi-hyperplanes Π_0 and Π_φ is equal to φ.

Proof. The space R^{2n} should be considered as $2n$-dimensional Euclidian space with the obvious scalar product $(a,b) = \sum_{j=1}^{2n} a_j b_j$, where $a = (a_1,...,a_{2n})$, $b = (b_1,...,b_{2n})$. Therefore cosine of angle γ between the vectors $a, b \in R^{2n}$ may be evaluated as follows:

$$\cos\gamma = \frac{(a,b)}{|a\|b|}, \qquad (2.3.14)$$

where $|a| = \sqrt{(a,a)}$ is a length of the vector a. The angle between the semi-hyperplanes Π_0 and Π_φ is equal to the angle between the normal vectors $\vec{n}_0$ and $\vec{n}_\varphi$.

Let $\vec{n}_0 = (a_1,...,a_{2n})$ and $\vec{n}_\varphi = (b_1,...,b_{2n})$. It follows from (2.3.5) that

$$a_1 = u_1 \sin 0 - v_1 \cos 0 = -v_1,$$

$$\dots\dots\dots\dots\dots\dots\dots\dots$$

$$a_n = u_n \sin 0 - v_n \cos 0 = -v_n,$$
$$a_{n+1} = -u_1 \cos 0 - v_1 \sin 0 = -u_1, \qquad (2.3.15)$$

$$\dots\dots\dots\dots\dots\dots\dots\dots$$

$$a_{2n} = -u_n \cos 0 - v_n \sin 0 = -u_n;$$
$$b_1 = u_1 \sin\varphi - v_1 \cos\varphi = -v_1,$$

$$\dots\dots\dots\dots\dots\dots\dots\dots$$

$$b_n = u_n \sin\varphi - v_n \cos\varphi = -v_n,$$
$$b_{n+1} = -u_1 \cos\varphi - v_1 \sin\varphi = -u_1, \qquad (2.3.16)$$

$$\dots\dots\dots\dots\dots\dots\dots\dots$$

$$b_{2n} = -u_n \cos\varphi - v_n \sin\varphi = -u_n.$$

Then taking into account the equation (2.3.14), we obtain the following:
$\cos\gamma = \dfrac{(\vec{n}_0,\vec{n}_\varphi)}{|\vec{n}_0\|\vec{n}_\varphi|}$. Let us substitute the coordinates of the vectors $\vec{n}_0$ and $\vec{n}_\varphi$ from the equations (2.3.15 - 2.3.16) in the last equation. Then we obtain the following:

$$\cos\gamma = \frac{\sum_{j=1}^{n}\left[(-v_j)(u_j\sin\varphi - v_j\cos\varphi) + (-u_j)(-u_j\cos\varphi - v_j\sin\varphi)\right]}{\sqrt{\sum_{j=1}^{n}(v_j^2 + u_j^2)}\sqrt{\left[(u_j\sin\varphi - v_j\cos\varphi)^2 + (-u_j\cos\varphi - v_j\sin\varphi)^2\right]}},$$

and after simplification:

$$\cos\gamma = \frac{\cos\varphi\sum_{j=1}^{n}(v_j^2 + u_j^2)}{\sqrt{\sum_{j=1}^{n}(v_j^2 + u_j^2)}\sqrt{\sum_{j=1}^{n}\left[u_j^2(\sin^2\varphi + \cos^2\varphi) + v_j^2(\cos^2\varphi + \sin^2\varphi)\right]}} = \cos\varphi.$$

Therefore the angle between Π_0 and Π_φ is equal to φ.

The following statement, which is more general than Theorem 2.3.6, is useful.

Theorem 2.3.7. The angle between semi-hyperplanes Π_φ and Π_ψ is equal to $\psi - \varphi$.

Proof. A proof of this theorem is similar to the proof of the previous one. It means that it is necessary to apply the formula (2.3.14) to the vectors $\vec{n}_\varphi$ and $\vec{n}_\psi$. Let

$$Q_j = \bigcup_{j \le \frac{k\varphi}{2\pi} < j+1} \Pi_\varphi, \quad j = 0,1,...,k-1. \qquad (2.3.17)$$

Taking into account that S^0 is a two-dimensional plane, we can conclude that $S^0 \cap Q_j$, $j = 0,1,...,k-1$ is a semi-open sector with the center M^0, and boundaried by rays $L_{\frac{2\pi}{k}j}$ and $L_{\frac{2\pi}{k}(j+1)}$ within the plane S^0. The first of these rays belongs to the noted sector (without the point M^0), and the second one does not belong to this sector. The angle between these sectors is equal to $\frac{2\pi}{\kappa}$ according to the Theorem 2.3.7.

Definition 2.3.2. An ordered collection of the sets

$$Q = \left\{Q_0, Q_1, ..., Q_{k-1}\right\} \qquad (2.3.18)$$

is a *k-edge* corresponding to the given linear function $W(X) = w_0 + w_1x_1 + ... + w_nx_n$. The set Q_j is a jth edge of the *k-edge* Q. The set T_0 (a plane of a dimension 2n-2) is a *sharp point* of the *k-edge* Q.

It should be noted that the edges $Q_0, Q_1, ..., Q_{k-1}$ of the k-edge Q are numbered by such a way that their intersections $Q_j \cap S^0$, $j = 0,1,...,k-1$ with the plane S^0 induce a sequence

$$(Q_0 \cap S^0, \ Q_1 \cap S^0, \ ..., \ Q_{k-1} \cap S^0) \qquad (2.3.19)$$

of the sectors of the same angle value $\dfrac{2\pi}{k}$ within the two-dimensional plane S^0, and following one by one in a positive direction. We assume that the positive direction is the direction of the rotation around the center M^0 within the plane S^0, from the ray L_0 to the ray $L_{\frac{2\pi}{k}}$.

Definition 2.3.3. An *ordered decomposition* of the non-empty set M is a sequence $M_1,...,M_s$ of non-empty subsets of the set M, which are mutually disjoint, and union of which is equal to the set M.

Let us denote an ordered decomposition via $[M_1,...,M_s]$, and the fact that it is an exact decomposition of the set M will be written as $M = [M_1,...,M_s]$.

$W(\alpha) = 0$ for $\alpha \in T_0$ according to the equations (2.3.7). As it was mentioned above, we have to put some value from the set $\widetilde{E}_k^n$ as a value of the function P on the zero ($P(0)$). It will be natural to connect the sharp point of the T_0-edge with the fixed edge Q_t. We can denote in such a case $Q_t \cap S_0 = \overline{Q}_t$. An ordered collection of the sets $\{Q_0, Q_1, ..., Q_t, ..., Q_{k-1}\}$ may be denoted as $\overline{Q}_t$, or $\overline{Q}$, if a number t is unknown, or, if its precise value is not important. In such a case $\overline{Q}$ also as Q is a k-edge corresponding to a linear function $W(X)$. The notations $Q = \{Q_0, Q_1, ..., Q_t, ..., Q_{k-1}\}$ and $\overline{Q} = \{Q_0, Q_1, ..., \overline{Q}_t, ..., Q_{k-1}\}$ are equivalent in such a case. In other words we will denote sometimes the component $Q_t \cap T_0$ of the k-edge $\overline{Q}_t$ as Q_t. Therefore one of the k-edges Q or $\overline{Q}$ shows, is Q_t connected with one of them. We can conclude now that $\overline{Q}$ forms an ordered decomposition of the space C^n:

$$C^n = [Q_0, Q_1,..., \overline{Q}_t,..., Q_{k-1}]. \qquad (2.3.20)$$

Theorem 2.3.8. If the vector $\alpha = (\alpha_1,...,a_n)$ from the space C^n belongs to the jth edge of the k-edge $\overline{Q}$, then $P(W(\alpha)) = \varepsilon^j$.

Proof. $\forall \alpha \in \Pi_\varphi \; W(\alpha) = \varphi$ according to the equations (2.3.7). It follows from the (2.3.17) that the following condition is true for $\Pi_\varphi \in Q_j$:

$$2\pi j / k \le \varphi < 2\pi(j+1) / k.$$

Therefore we obtain the following for $\alpha \in Q_j$:

$2\pi j / k \le \arg(W(\alpha)) < 2\pi(j+1)/k$. If $\alpha \in T_0$ (it also means that $\alpha \in \bar{Q}_t$) then $W(\alpha) = 0$, and therefore $P(W(\alpha)) = \varepsilon^t$, which completes the proof.

Theorem 2.3.9. If the condition $P(W(\alpha)) = \varepsilon^j$ is true for some $\alpha = (\alpha_1, \ldots, a_n) \in C^n$, then $\alpha \in Q_j$.

Proof. Let $W(\alpha) \ne 0$, and $P(W(\alpha)) = \varepsilon^j$. Then $2\pi j / k \le \arg(W(\alpha)) < 2\pi(j+1)/k$. The edges $Q_0, Q_1, \ldots, Q_{k-1}$ form an ordered decomposition of the space C^n, and therefore α belongs to one of the edges, and to one of the semi-hyperplanes Π_φ , exactly to such that $2\pi j / k \le \varphi < 2\pi(j+1)/k$. It means that $\alpha \in Q_j$ according to (2.3.2). Let $P(W(\alpha)) = \varepsilon^t$ now, and $W(\alpha) = 0$. Then $\alpha \in T_0$ according to (2.3.7), and therefore also $\alpha \in \bar{Q}_t$.

We are ready now to consider a very important correspondance between the notions of the k-valued threshold function and k-edge.

Theorem 2.3.10. Let $f(x_1, \ldots x_n)$ be a k-valued threshold function, which is defined on the set $Z \subset C^n$, and at least one of the following conditions is true: 1) The set Z is bounded; 2) $P(f(x_1, \ldots x_n)) \ne const$. Then such a k-edge $\bar{Q} = \{Q_0, Q_1, \ldots, Q_{k-1}\}$ exists that

$$\forall \, \alpha = (\alpha_1, \ldots, \alpha_n) \in Z \cap Q_j \; P(f(\alpha)) = \varepsilon^j.$$

Proof. Let $W = (w_0, w_1, \ldots, w_n)$ be such a weighting vector of the given function f that at least one of its components $w_1, \ldots, w_n$ is not equal to zero. According to the conditions of the theorem and to the above considerations it is always possible to find such a weighting vector for function f.

Let us consider a linear function $W(X) = w_0 + w_1 x_1 + \ldots + w_n x_n$. Let $\bar{Q} = \{Q_0, Q_1, \ldots, Q_{k-1}\}$ be k-edge corresponding to this linear function. Since f is a threshold function, then $P(f(X)) = P(W(X))$. Therefore if $X \in Z \cap Q_j$ then according to the Theorem 2.3.8 $P(f(\alpha)) = \varepsilon^j$. Theorem is proven.

The inverse theorem is also true. Moreover, it is not necessary to put the restrictions on the domain of function f.

Theorem 2.3.11. Let $f(x_1,...x_n)$ be a complex-valued function, which is defined on the set $Z \subset C^n$, and assume that a k-edge $\overline{Q} = \{Q_0, Q_1, ..., Q_{k-1}\}$ exists such that $\forall \alpha = (\alpha_1,...,\alpha_n) \in Z \cap Q_j \ P(f(\alpha)) = \varepsilon^j$. Then $f(X)$ is a k-valued threshold function.

Proof. Let $W(X) = w_0 + w_1 x_1 + ... + w_n x_n$ be a linear function, to which k-edge $\overline{Q}$ corresponds. If $\alpha \in Z$, and $P(f(\alpha)) = \varepsilon^j$, then according to the Theorem 2.3.9 $\alpha \in Q_j$. Therefore $\forall \alpha \in Z \ P(f(\alpha)) = P(W(\alpha))$, and it means that the function $f(x_1,...x_n)$ is threshold.

Definition 2.3.4. A *k-edge of the function f* is such a k-edge $\overline{Q}$, which satisfies the conditions of the Theorem 2.3.11.

Let $f(x_1,...x_n)$ is an arbitrary complex-valued function of the complex-valued variables which is defined on the set $Z \subset C^n$. Let us denote

$$A_j = \{\alpha \in Z | P(f(\alpha)) = \varepsilon^j\}, \quad j = 0, 1, .., k\text{-}1. \tag{2.3.21}$$

Certainly, some of the sets A_j may be empty. Despite this fact it will be convenient to use the term "an ordered decomposition", when the sets A_j, $j = 0, 1, ..., k$-1 are used.

Definition 2.3.5. An ordered collection of the sets $A_0, A_1, ..., A_{k-1}$ (see (2.3.21)) is called a *P-decomposition* of the domain Z of the function $f(x_1,...x_n)$.

P-decomposition will be denoted by symbol $\left[A_0, A_1, ..., A_{k-1}\right]$ and similar to the ordered decomposition of the arbitrary set M (see the Definition 2.3.3 above) we will write that $Z = \left[A_0, A_1, ..., A_{k-1}\right]$. The elements $A_0, A_1, ..., A_{k-1}$ of the P-decomposition are the complete prototypes of the values of function $P(f(X))$, which is defined on the set Z.

The following theorem is very important for understanding the geometrical meaning of the notion of k-valued threshold function.

Theorem 2.3.12. The complex-valued function $f(x_1,...x_n)$, which is defined on the set $Z \subset C^n$, is a k-valued threshold function if and only if a k-edge $\overline{Q} = \{Q_0, Q_1, ..., Q_{k-1}\}$ exists such that its elements (edges) are connected with the sets $A_0, A_1, ..., A_{k-1}$ of the P-decomposition $Z = \begin{bmatrix} A_0, A_1, ..., A_{k-1} \end{bmatrix}$ by following relation:

$$A_j \subset Q_j, \quad j = 0, 1, ..., k-1. \tag{2.3.22}$$

Proof. Necessity. Let the function $f(X)$ be threshold, and $W = (w_0, w_1, ..., w_n)$ is such a weighting vector of the function f that $w_t \neq 0$ at least for some $t \in \{1, 2, ..., n\}$. Let also $\overline{Q} = \{Q_0, Q_1, ..., Q_{k-1}\}$ be a k-edge corresponding to this linear function $W(X)$. Then according to the Theorem 2.3.9 and Definition 2.3.5 (see formula (2.3.21) also) we obtain $A_t \subset Q_t$.

Sufficiency. Let $A_t \subset Q_t$ for all $t \in \{0, 1, ..., k-1\}$, and $W(X) = w_0 + w_1 x_1 + ... + w_n x_n$ be a linear function. Let $\overline{Q} = \{Q_0, Q_1, ..., Q_{k-1}\}$ be a k-edge corresponding to this linear function. Then according to the Theorem 2.3.8 the condition $P(W(\alpha)) = \varepsilon^t$ is true for $\alpha \in A_t \subset Q_t$. On the other hand it follows from the (2.3.21) that $P(f(\alpha)) = \varepsilon^t$. So, $\forall \alpha \in Z \ P(f(\alpha)) = P(W(\alpha))$, and it means that $f(X)$ is a k-valued threshold function with the weighting vector W.

Definition 2.3.6. The sets $A_0, A_1, ..., A_{k-1}$ establish *edge-like sequence*, if such a k-edge $\overline{Q} = \{Q_0, Q_1, ..., Q_{k-1}\}$ exists, that $A_j \subset Q_j$ for all $j \in \{0, 1, ..., k-1\}$. In such a case the ordered decomposition $\begin{bmatrix} A_0, A_1, ..., A_{k-1} \end{bmatrix}$ of some set $A = A_0 \cup A_1 \cup ... \cup A_{k-1} \subset C^n$ is *edge-like decomposition*.

The following theorem follows directly from the Theorem 2.3.12 and Definitions 2.3.5 and 2.3.6.

Theorem 2.3.13. The complex-valued function $f(x_1, ... x_n)$, which is defined on the set $Z \subset C^n$, is a k-valued threshold function if and only if a P-decomposition of the set Z is an edge-like decomposition.

Let clarify a geometrical meaning of the k-edge corresponding to the linear function $w_0 + w_1 x$ ($w_1 \neq 0$) of a single complex variable e.g., threshold function $f(X)$ of the single complex variable x (see the Definition 2.3.4).

In such a case ($n=1$) the sequence $(Q_0 \cap S^0, Q_1 \cap S^0, ..., Q_{k-1} \cap S^0)$ (see (2.3.19)) coincides with the k-edge $Q = \{Q_0, Q_1, ..., Q_{k-1}\}$ (see the Definition 2.3.2), and the set T_0, which is a sharp point of the k-edge is a plane of dimension $2n - 2 = 2 \cdot 1 - 2 = 0$. Therefore T_0 is a point, and it coincides with the point M^0. Thus, k-edge Q for $n=1$ is a sequence of the sectors on the obvious complex plane $C = S^0$, which are created by k rays, with the same angle value $2\pi / k$ and the same center M^0. These rays follow each other in the positive direction (against the clock arrow), and each sector contains only the first of two rays, which are its boundaries.

After connection of the sharp-point M^0 with the fixed edge Q_t we obtain the k-edge $\overline{Q} = \{Q_0, Q_1, ..., \overline{Q_t}, ..., Q_{k-1}\}$ corresponding to the linear function $w_0 + w_1 x$ on the plane $C = S^0$. According to (2.3.21) it is possible to say that edges of the k-edge establish the ordered decomposition of the all the points of the complex plane

$$C = [Q_0, Q_1, ..., Q_t, ..., Q_{k-1}]. \qquad (2.3.23)$$

To complete consideration of the interpretation of k-valued threshold functions and multi-valued neurons we have to consider the following definition of the k-valued threshold function, which is generalization of the Definition 2.2.1.

Definition 2.3.7. The complex-valued function $f(x_1,...,x_n)$, which is defined on the set $Z \subset C^n$ is called a *k-valued threshold function*, if it is possible to find such a complex-valued weighting vector $W = (w_0, w_1,..., w_n)$, and to define $P(0)$ in such a way that the equation (2.3.1)

$$P\big(f(x_1,...x_n)\big) = P(w_0 + w_1 x_1 + ... + w_n x_n)$$

(where P is the function (2.2.2)) will be true for all values of the variables x from the domain of the function f.

The Definitions 2.2.1 and 2.3.7 of k-valued threshold function differ from each other in the following aspect: the Definition 2.2.1 uses a function P, which is not defined in $(0,0)$, and the Definition 2.3.7 uses a function P which is defined in $(0,0)$. Both definitions will be equivalent if they define the same classes of the k-valued threshold functions.

Theorem 2.3.14. Equivalence of the first and second definitions of the k-valued threshold function $f(x_1,...x_n)$, which is defined on the set $Z \subset C^n$, does not depend on a mean of the definition the function P in $(0,0)$ ($P(0)$) if and only if a weighting vector $W = (w_0, w_1, ..., w_n)$ exists for the threshold function f (according to the Definition 2.2.1) such that the set $\{\alpha \in C^n | W(\alpha) = 0\}$ does not intersect with the domain Z of the function f:

$$Z \cap \{\alpha \in C^n | W(\alpha) = 0\} = \varnothing. \qquad (2.3.24)$$

Proof. A sufficiency of the condition (2.3.24) is evident because according to the Definition 2.2.1 $\forall \alpha \in Z \; P(f(\alpha)) = P(W(\alpha))$, and therefore taking into account (2.3.24): $\forall \alpha \in Z \; W(\alpha) \neq 0$.

Necessity. Both definitions of k-valued threshold function are equivalent independently of the definition the $P(0)$. Let us assume that a weighting vector $W = (w_0, w_1, ..., w_n)$ exists for the threshold function f (according to the Definition 2.3.7) such that $Z \cap \{\alpha \in C^n | W(\alpha) = 0\} = A \neq \varnothing$. It means that for any $\alpha \in A$ $P(W(\alpha))$ will take some specific values, such that $P(W(\alpha)) = P(0)$ has to be equal. But it contradicts to arbitrariness of the definition the $P(z)$ in the $(0,0)$.

It follows from the Theorem 2.3.14 that it is possible to omit a definition of the $P(z)$ in the $(0,0)$ or it is possible to define it arbitrarily if the condition (2.3.24) is true.

Finally, let us consider the following generalization of the notion of k-valued threshold function. $C \setminus \{0\} = [s_0, s_1, ..., s_{k-1}]$ is an ordered decomposition of the set of points on the complex plane (without point $(0, 0)$) onto mutual disjoint sectors $s_0, s_1, ..., s_{k-1}$, which are bounded by rays $l_0, l_1, ..., l_{k-1}$ started from the point $(0,0)$. Only the first of two rays, which form a sector, is included to it. The direction of the ray l_0 coincides with the direction of the positive real semi-axis. The angle values $\varphi_0, \varphi_1, ..., \varphi_{k-1}$ of the sectors $s_0, s_1, ..., s_{k-1}$ are arbitrary and must satisfy the only condition: $\varphi_0 + \varphi_1 + ... + \varphi_{k-1} = 2\pi$. Let

$$\widetilde{P}(z) = \varepsilon^j, \text{if } z \in s_j, \; j = 0, 1, ..., k-1, \; \varepsilon = 2\pi i / k. \qquad (2.3.25)$$

It is possible to consider a definition of the multiple-valued threshold function and multi-valued neuron, which coincides, with the Definitions 2.2.1 and 2.3.1, respectively with the only difference: instead of the function P defined by (2.2.2), the function $\widetilde{P}$ defined by (2.3.25) will be used. The function P differs from the function $\widetilde{P}$ by angle value of the sectors, to which the corresponding function separates the complex plane. All the sectors for the function P are equal each other, and their angle values are equal exactly to $2\pi / k$. The sectors corresponding to function $\widetilde{P}$ may have arbitrary angle value. We will not return here to the multiple-valued functions and multi-valued neurons defined using (2.3.25), but from our point of view such a generalization may be very interesting from the point of view of increasing the MVN's functionality. It will be very important to return to this generalization in the further work.

4. SYNTHESIS OF THE MULTI-VALUED NEURON USING LINEAR PROGRAMMING METHOD

A problem of synthesis of the multi-valued neuron is a problem of finding of a weighting vector $W = (w_0, w_1, ..., w_n)$, which implements input/output mapping described by the multiple-valued function f.

This problem should be divided on two sub-problems: 1) to check if the function f is a threshold function for a fixed value of k; 2) If yes, search at least for the one weighting vector implementing the function f.

The second of these sub-problems may be solved by learning. In Chapter 4 we will consider in detail effective learning algorithms for MVN with two different learning rules. But it is also possible to solve both of sub-problems simultaneously using properties of group characters considered within the Section 1.1 and linear programming methods. More exactly using properties of the group characters it is possible to reduce the problem of the MVN synthesis to the solution of the linear programming problem. We will consider this approach here.

Let us consider first of all the MVN synthesis for the mappings described by fully defined k-valued functions. Then synthesis for the mappings described by partially defined k-valued functions $(k>2)$ will be also considered.

It should be remainded that the vectors $X_1,...,X_n$ of values of the k-valued variables $x_1,...,x_n$ and the vector $X_0 = (1,...,1)$ are characters of the group E_k^n with the numbers 0, k^{n-j-1}, $j=1, ..., n$. These vectors simultaneously completely coincide with the generalized Rademacher functions (see Section 2.2 above). Let us renumber the characters of the group E_k^n as follows: $\chi_0 = X_0, \chi_1 = X_1, ..., \chi_n = X_n, \chi_{n+1}, ..., \chi_{k^n-1}$, so $\chi_{n+1}, ..., \chi_{k^n-1}$ are all the characters, which are different from the generalized Rademacher functions.

Theorem 2.4.1. The function $f(X)$ of the k-valued logic is threshold if and only if such a function $F(X): \tilde{E}_k^n \to C$, $(F(X) \neq 0$ for any X) exists that

$$\forall \alpha \in \tilde{E}_k^n \quad 0 \leq \arg(F(\alpha)) < \frac{2\pi}{k} \qquad (2.4.1)$$

and

$$\left(F(X)f(X), \chi_j(X)\right) = 0 \qquad (2.4.2)$$

for all the characters χ_j of the group $\tilde{E}_k^n$, which are not equal to the characters $X_0, X_1, ..., X_n$ of this group (i.e. the generalized Rademacher functions).

Proof. Necessity. Let $f(X)$ be a threshold function with the weighting vector $W = (w_0, w_1, ..., w_n)$. According to the Theorem 2.2.1 a function $F(X)$, which is defined on $\tilde{E}_k^n$, and takes non-zero values in the field of the complex numbers, exists such that

$$F(X)f(X) = w_0 + w_1 x_1 + ... + w_n x_n \qquad (2.4.3)$$

or

$$F(X)f(X) = (W, X) \qquad (2.4.4)$$

and

$$\forall \alpha \in \tilde{E}_k^n \quad 0 \le \arg(F(\alpha)) < \frac{2\pi}{k}. \qquad (2.4.5)$$

Let us decompose the function $F(X)f(X)$ with respect to the basis of the unitary space U, which consists of the characters of the group $\tilde{E}_k^n$:

$$F(X)f(X) = \gamma_0 x_0 + \gamma_1 x_1 + ... + \gamma_n x_n + \sum_{j=n+1}^{k^n-1} \gamma_j \chi_j. \qquad (2.4.6)$$

The characters χ_j under the sign of $\sum$ in the equality (2.4.6) are not equal to the generalized Rademacher functions. There are exactly $l = k^n - 1$ of such characters. But the decomposition (2.4.6) is unique, hence it means that

$$\gamma_0 = w_0, \gamma_1 = w_1, ..., \gamma_n = w_n, \gamma_{n+1} = \gamma_{n+2} = ... = \gamma_l = 0, \qquad (2.4.7)$$

It is evident from the (2.4.7) that

$$\forall j \in \{n+1, n+2, ..., k^n - 1\} \quad \gamma_j = \frac{1}{k^n}(F(X)f(X), \chi_j(X)) = 0, \quad \text{and}$$

all the characters χ_j are not equal to the characters $X_0, X_1, ..., X_n$ of the group $\tilde{E}_k^n$. It means that the function $F(X)$ satisfies the conditions of the theorem, and necessity is proven.

Sufficiency. Let $F(X)$ satisfy the conditions of the theorem. Let us decompose the function $F(X)f(X)$ in the basis of the unitary space U, which consists of the characters of the group $\tilde{E}_k^n$:

$$F(X)f(X) = \gamma_0 x_0 + \gamma_1 x_1 + ... + \gamma_n x_n + \sum_{j=n+1}^{k^n-1} \gamma_j \chi_j.$$

Since according to the condition of the theorem

$$\forall j \in \{n+1, n+2, ..., k^n - 1\} \quad \gamma_j = \frac{1}{k^n}(F(X)f(X), \chi_j(X)) = 0, \text{ then}$$

$$F(X)f(X) = w_0 + w_1 x_1 + ... + w_n x_n.$$

According to the Theorem 2.2.1 the function $f(X)$ is threshold, and the following equalities are true for its weighting vector $W = (w_0, w_1, ..., w_n)$:

$$w_j = \gamma_j = \gamma_j = \frac{1}{k^n}(F(X)f(X), \chi_j(X)), \quad j = 0, 1, ..., n.$$

The theorem is proven.

It should be noted that an arbitrary complex-valued function $F(X)$ may be presented as follows for $k \geq 3$: $F(X) = A(X) + \varepsilon B(X)$, where $\varepsilon = e^{i2\pi/k}$ is a primitive k^{th} root of unity, and $A(X)$, $B(X)$ are some real-valued functions. In fact, $\arg \varepsilon = 2\pi / k$, and the vector corresponding to the number ε on the complex plane is not collinear to the vector corresponding to real unity, when $k \geq 3$. This justifies the noted presentation.

It is possible to show that the complex-valued function $F(X) = A(X) + \varepsilon B(X) \neq 0$ satisfy the condition $0 \leq \arg(F(X)) < \dfrac{2\pi}{k}$ if, and only if, the real-valued functions $A(X)$, $B(X)$ satisfy the condition $A(X) > 0, B(X) \geq 0$. Taking the last fact into account we can formulate the Theorem 2.4.1 in following way.

Theorem 2.4.2. The function $f(X)$ of the k-valued logic (for $k \geq 3$) is threshold if and only if such real-valued functions $A(X) > 0, B(X) \geq 0$ exist that

$$\left([A(X) + \varepsilon B(X)]f(X), \chi_j\right) = 0, \qquad (2.4.8)$$

if $\chi_j \neq x_l$ for all the $l = 0, 1, 2, ..., n$.

Let the function $F(X)$ and the function of the k-valued logic $f(X)$ satisfy the conditions of the Theorem 2.2.5. We will say that the vector $W = (w_0, w_1, ..., w_n)$ corresponds to the pair (f, F), if

$$\forall j \in \{0, 1, ..., n\} \quad w_j = \frac{1}{k^n}(F(X)f(X), X_j).$$ Let $a_{i_1 i_2 ... i_n} = a_i$ be a value of the function $A(X)$ on $\alpha = (\varepsilon^{i_1}, \varepsilon^{i_2}, ..., \varepsilon^{i_n})$, where $i = i_1 k^{n-1} + ... + i_{n-1} k + i_n$, and b_i is a value of the function $B(X)$ defined by similar way.

Let us consider the following example to illustrate the Theorems 2.4.1 - 2.4.2 and equations (2.4.7)-(2.4.8). Let $k=3$, $n=1$, $f(1) = f(\varepsilon)=1$, $f(\varepsilon^2) = \varepsilon$. Table 2.4.1 contains the values of the characters and functions $f(X)$, $A(X)$, $B(X)$.

Table 2.4.1 Example for illustration of the Theorems 2.4.1 – 2.4.2

X	X_0	$X_1 = X$	$X_2 = X^2$	$f(X)$	$A(X)$	$B(X)$
1	1	1	1	1	a_0	b_0
ε	1	ε	ε^2	1	a_1	b_1
ε^2	1	ε^2	ε	ε	a_2	b_2

Only the character X_2 of the group $\widetilde{E}_3^1$ satisfies the condition $X_j \neq x_l$ for $l = 0, 1$, and the system (2.4.8) contains the only equation

$$([A(X) + \varepsilon B(X)]f(X), X^2) = 0, \qquad (2.4.9)$$

or after transformation:

$$(A(X)f(X), X^2) + \varepsilon(B(X)f(X), X^2) = 0. \qquad (2.4.10)$$

The equation (2.4.10) is transformed to

$$a_0 + a_1\varepsilon^2 + a_2\varepsilon^2 + b_0\varepsilon + b_1 + b_2 = 0.$$

The last equation has the following solution: $a_0 = a_1 = a_2 = b_1 = 1$, $b_0 = 2$, $b_2 = 0$, which satisfy the conditions $A(X) > 0, B(X) \geq 0$ and the equations (2.4.9)-(2.4.10). It means that function $f(X)$ is threshold. Let us find its weighting vector:

$$w_0 = \frac{1}{3}([A(X) + \varepsilon B(X)]f(X), X_0) = \frac{2 + 4\varepsilon}{3};$$

$$w_1 = \frac{1}{3}([A(X) + \varepsilon B(X)]f(X), X) = \frac{\varepsilon^2 + 3\varepsilon + 2}{3};$$

$$W = \left(\frac{2 + 4\varepsilon}{3}, \frac{\varepsilon^2 + 3\varepsilon + 2}{3}\right).$$

Taking into account that $\forall c > 0$ $P(cX) = P(X)$, we can multiply all the weights by 3, and finally obtain the weighting vector $W' = 3W = (2 + 4\varepsilon, \varepsilon^2 + 3\varepsilon + 2)$.

We can to conclude that a problem of the synthesis the MVN which implements the mapping described by k-valued function $f(X)$ is reduced for $k > 2$ to search for a real-valued solution $a = (a_0', a_1', ..., a_l', b_0', b_1', ..., b_l')$ of system of the linear algebraic homogeneous equations (2.4.8) with complex-valued coefficients. This solution has to satisfy the condition

$$a_j' > 0, \ b_j' \geq 0; \ j = 0, 1, ..., l. \qquad (2.4.11)$$

Let us solve for the beginning the same problem for arbitrary system of linear algebraic homogeneous equations with real-valued coefficients. Let

$q_i = (q_{i_1}, q_{i_2}, ..., q_{i_l}), j = 1, 2, ..., l$ are sets of real-valued coefficients, and γ is a vector of unknowns. We have to find such a solution $\gamma = (\gamma_1, \gamma_2, ..., \gamma_t, \gamma_{t+1}, \gamma_l)$ of the linear homogeneous system

$$(q_i\gamma) = \sum_{j=1}^{l} q_{ij}\gamma_j = 0, \quad i = 1, 2, ..., r; \quad r \le l, \tag{2.4.12}$$

which satisfies the conditions

$$\begin{aligned} \gamma_j > 0, \quad j = 1, 2, ..., t \\ \gamma_j \ge 0, \quad j = t+1, ..., m. \end{aligned} \tag{2.4.13}$$

Theorem 2.4.3. The system of the linear homogeneous algebraic equations (2.4.12) has a solution $\gamma = (\gamma_1, \gamma_2, ..., \gamma_t, \gamma_{t+1}, \gamma_l)$ which satisfies the condition (2.4.13) if and only if a non-negative solution exists for the following system

$$(q_i\gamma) = -\sum_{j=1}^{t} q_{ij} \quad i = 1, 2, ..., r. \tag{2.4.14}$$

Proof. Sufficiency. Let $\gamma' = (\gamma_1', \gamma_2', ..., \gamma_t', \gamma_{t+1}', \gamma_l')$ be a non-negative solution of the system (2.4.14). Let s be l-dimensional vector, first t components of which are units, and all the others are zeros. Let $\gamma = \gamma' + s$. Therefore $\gamma_i > 0$, $i = 1,2,...,t$, and $\gamma_i \ge 0$, $i = t+1, t+2, ..., l$. In such a case $(q_i, \gamma) = -\sum_{i=1}^{t} q_{ij} + \sum_{i=1}^{t} q_{ij} = 0$, and γ is a solution of the system (2.4.12).

Necessity. Let $\gamma'' = (\gamma_1'', \gamma_2'', ..., \gamma_t'', \gamma_{t+1}'', \gamma_l'')$ be a solution of the system (2.4.12), which satisfies the condition (2.4.13). Let us choose some $c > 0$ in such a way that the condition $c\gamma_i'' \ge 1$ will be true for all $i = 1, 2, ..., t$. Let $\gamma' = c\gamma'' - s$. It means that the components of the vector γ' are non-negative, and $(q_i, \gamma') = c(q_i, \gamma'') - (q_i, s)$. Since γ'' is a solution of the system (2.4.12), then $(q_i, \gamma') = -(q_{ij}, s) = -\sum_{i=1}^{t} q_{ij}$, and γ' is a solution of the system (2.4.14). Theorem is proven.

Let $q_i = (q_{i_1}, q_{i_2}, ..., q_{i_l}), j = 1, 2, ..., l$ are sets of the complex-valued coefficients. It is possible to reduce a search for the real-valued solution of the system of linear homogeneous equations (2.4.12) with the complex-valued coefficients, which satisfies the condition (2.4.13), to just described solution of the system with real-valued coefficients. Really, the system (2.4.12) with the complex-valued coefficients has a real-valued solution γ, which satisfies the condition (2.4.13) if and only if the following system has a real-valued solution:

$$\sum_{j=1}^{l} \mathrm{Im}\, q_{ij} \gamma_i = 0,$$

$$\sum_{j=1}^{l} \mathrm{Re}\, q_{ij} \gamma_i = 0. \tag{2.4.15}$$

The simplex-method for the linear programming problem [Dantzig (1963)] can be used to search for a non-negative solution of the system of equations with the real-valued coefficients.

$$a_{11}x_1 + a_{12}x_2 + ... + a_{1n}x_n = b_1,$$

$$...$$

$$a_{i1}x_1 + a_{i2}x_2 + ... + a_{in}x_n = b_i, \tag{2.4.16}$$

$$...$$

$$a_{r1}x_1 + a_{r2}x_2 + ... + a_{mn}x_n = b_r.$$

To solve a problem with such an approach it is sufficient to find a minimum of the function $h = x_{n+1} + x_{n+2} + ... + x_{n+r}$ on the set of non-negative solutions of the system

$$a_{11}x_1 + a_{12}x_2 + ... + a_{1n}x_n + x_{n+1} = b_1,$$

$$...$$

$$a_{i1}x_1 + a_{i2}x_2 + ... + a_{in}x_n + x_{n+2} = b_i, \tag{2.4.17}$$

$$...$$

$$a_{r1}x_1 + a_{r2}x_2 + ... + a_{mn}x_n + x_{n+r} = b_r.$$

The system (2.4.17) has non-negative solutions if and only if $\min h = 0$. If this condition is gotten on the solution $\gamma = (\gamma_1, \gamma_2, ..., \gamma_n, ..., \gamma_{n+r})$ of the system (2.4.17) then $\gamma_{n+1} = \gamma_{n+2} = ... = \gamma_{n+r} = 0$ and $\gamma' = (\gamma_1, \gamma_2, ..., \gamma_n)$ is a non-negative solution of the system (2.4.16). Function $f(X)$ in such a case is threshold. If $\min h > 0$ then system (2.4.16) has no non-negative solutions, and function $f(X)$ is not threshold.

Let us consider two examples, which will illustrate the different situations that may appear in context of using the method of MVN synthesis.

Example 2.4.1. Let a mapping implemented by MVN with the single input is described by following function of the four-valued logic: $f(1) = 1, f(i) = f(-1) = i, f(-i) = -i$ (i is imaginary unity). Table 2.4.2 contains the corresponding values of characters, functions $f(X)$, $A(X)$, and $B(X)$. The system (2.4.8) contains 4-2=2 equations:

$$\big([A(X)+iB(X)]f(X),X^2\big)=a_0-a_1i+a_2i+a_3i+b_0i+b_1-b_2-b_3=0,$$
$$\big([A(X)+iB(X)]f(X),X^3\big)=a_0-a_1-a_2i-a_3+b_0i-b_1i-b_2-b_3i=0.$$

Table 2.4.2. Illustration to Example 2.4.1.

X_0	$X_1=X$	$X_2=X^2$	$X_3=X^3$	$f(X)$	$A(X)$	$B(X)$
1	1	1	1	1	a_0	b_0
1	i	-1	$-i$	i	a_1	b_1
1	-1	1	-1	i	a_2	b_2
1	$-i$	-1	i	$-i$	a_2	b_2

The system (2.4.15) is the following:

$$\begin{aligned}
a_0+b_1-b_2-b_3&=0,\\
-a_1+a_2+a_3+b_0&=0,\\
a_0-a_1-a_3+b_2&=0,\\
-a_2+b_0-b_1-b_3&=0.
\end{aligned}$$

(2.4.18)

We have to find a solution of the system (2.4.18), which satisfies the condition $a_j>0$, $b_j\geq0$, $j=0$, 1, 2, 3. This solution exists if and only if it is possible to find a non-negative solution for the system

$$\begin{aligned}
a_0+b_1-b_2-b_3&=-(1+0+0+0),\\
-a_1+a_2+a_3+b_0&=-(0-1+1+1),\\
a_0-a_1-a_3+b_2&=-(1-1+0-1),\\
-a_2+b_0-b_1-b_3&=-(0+0-1+0).
\end{aligned}$$

(2.4.19)

The following system is equivalent to the system (2.4.19):

$$\begin{aligned}
-a_0-b_1+b_2+b_3&=1,\\
a_1-a_2-a_3-b_0&=1,\\
a_0-a_1-a_3+b_2&=1,\\
-a_2+b_0-b_1-b_3&=1.
\end{aligned}$$

(2.4.20)

The matrix of the last system

$$\tilde{B}=\begin{pmatrix}
-1 & 0 & 0 & 0 & 0 & 0 & -1 & 1 & 1 & 1\\
0 & 1 & -1 & -1 & -1 & 0 & 0 & 0 & 0 & 1\\
1 & -1 & 0 & -1 & 0 & 0 & 1 & 0 & 1\\
0 & 0 & -1 & 0 & 1 & -1 & 0 & -1 & 1
\end{pmatrix}.$$

The extended matrix of the system (2.4.17) is the following:

$$\tilde{A} = \begin{pmatrix} -1 & 0 & 0 & 0 & 0 & -1 & 1 & 1 & 1 & 0 & 0 & 0 & 1 \\ 0 & 1 & -1 & -1 & -1 & 0 & 0 & 0 & 0 & 1 & 0 & 0 & 1 \\ 1 & -1 & 0 & -1 & 0 & 0 & 1 & 0 & 0 & 0 & 1 & 0 & 1 \\ 0 & 0 & -1 & 0 & 1 & -1 & 0 & -1 & 0 & 0 & 0 & 1 & 1 \\ 0 & 0 & -2 & -2 & 0 & -2 & 2 & 0 & 0 & 0 & 0 & 0 & 4 \end{pmatrix}.$$

Let us apply a simplex-algorithm of the linear programming. The simple transformations reduce $\tilde{A}$ to the following matrix:

$$\tilde{A} \sim \begin{pmatrix} 0 & 1 & -2 & -1 & 0 & -1 & 0 & -1 & 0 & 0 & 0 & 1 & 2 \\ 2 & 0 & -2 & -2 & 0 & 0 & 0 & -2 & -1 & 1 & 1 & 1 & 2 \\ 0 & 0 & -1 & -1 & 0 & -1 & 1 & 0 & 0.5 & 0.5 & 0.5 & 0.5 & 2 \\ 0 & 0 & -1 & 0 & 1 & 1 & 0 & -1 & 0 & 0 & 0 & 1 & 1 \\ 0 & 0 & -1 & 0 & 0 & 0 & 0 & 0 & -1 & -1 & -1 & -1 & 0 \end{pmatrix}.$$

The last row of the obtained matrix, which corresponds to the cost function, does not contain the positive numbers. It means that $\min h = 0$. The system (2.4.20) has a non-negative solution, and it means that the system (2.4.18) has a solution, which satisfies the condition $a_j > 0,\ b_j \geq 0,\ j = 0, 1, 2, 3$:

$$a_0 = 2,\ a_1 = 3,\ a_2 = 1,\ a_3 = 1,$$
$$b_0 = 1,\ b_1 = 0,\ b_2 = 0,\ b_3 = 1.$$

It means that function $f(X)$ is threshold, and its weighting vector can be found as follows:

$$w_0 = \frac{1}{4}\big([A(X) + iB(X)]f(X), \chi_0\big) = \frac{1}{4} \cdot 4i = i;$$

$$w_1 = \frac{1}{4}\big([A(X) + \varepsilon B(X)]f(X), X\big) = \frac{1}{4} \cdot 8 = 2;$$

$$W = (i, 2).$$

Example 2.4.2. Let a mapping implemented by MVN with the single input is described by following function of the four-valued logic: $f(1) = 1, f(i) = -i, f(-1) = 1, f(-i) = -1$ (i is an imaginary unit). The system (2.4.8) contains 4-2=2 equations:

$$\big([A(X) + iB(X)]f(X), X^2\big) = a_0 - a_1 i + a_2 + a_3 + b_0 i + b_1 + b_2 i + b_3 i = 0,$$

$$\big([A(X) + iB(X)]f(X), X^3\big) = a_0 - a_1 - a_2 + a_3 i + b_0 i - b_1 i - b_2 i - b_3 = 0.$$

The system (2.4.15) is the following:

$$a_0 + a_2 + a_3 + b_1 = 0,$$
$$-a_1 + b_0 + b_2 + b_3 = 0,$$
$$a_0 - a_1 - a_2 - b_3 = 0,$$
$$a_3 + b_0 - b_1 - b_2 = 0.$$

The first equation from this system has not solution satisfying the conditions $a_j > 0$, $b_j \geq 0$, $j = 0, 1, 2, 3$, and therefore function $f(X)$ is not a threshold function within the four-valued logic.

Let us consider a problem of the MVN synthesis for a mapping between neuron inputs and output, which is described by a partially defined multiple-valued function. Such functions precisely describe mappings corresponding to different applied problems, for example, problems of pattern recognition, classification, and time-series prediction.

What is a partially defined k-valued function? If $\widetilde{G}$ is a group, on which our k-valued logic is built, and some function $f(X)$ is a partially-defined function, it means that such a function is not defined on all the elements of the set $\widetilde{G}$. Let U is a subset of $\widetilde{G}$, on which function $f(X)$ is defined, U_1 is a subset of $\widetilde{G}$, on which function $f(X)$ is not defined, and $U \bigcup U_1 = \widetilde{G}$.

Theorem 2.4.4. The partially defined function $f(X)$ of the k-valued logic is a threshold function with weighting vector $W = (w_0, w_1, ..., w_n)$ if and only if such a complex-valued function $F(X) \neq 0$, which is defined on U, exists, that $\forall \alpha \in U$:

$$F(\alpha)f(\alpha) = W(\alpha), \tag{2.4.21}$$

$$0 \leq \arg(F(\alpha)) < \frac{2\pi}{k}. \tag{2.4.22}$$

The proof of the theorem is completely similar to the proof of the Theorem 2.2.1 for fully defined functions.

Let now consider the following function:

$$f_1(\alpha) = \begin{cases} f(\alpha), & \text{if } \alpha \in U \\ 1, & \text{if } \alpha \in U_1 \end{cases}.$$

Theorem 2.4.5. The partially-defined k-valued logic function $f(X)$ is a threshold function with weighting vector $W = (w_0, w_1, ..., w_n)$, if and only if a complex-valued function $F(X) \neq 0$, which is defined on $\widetilde{G}$, exists such that $\forall \alpha \in U$, $F(\alpha) \neq 0$:

$$F(\alpha)f_1(\alpha) = W(\alpha), \tag{2.4.23}$$

$$0 \leq \arg(F(\alpha)) < \frac{2\pi}{k}. \tag{2.4.24}$$

Proof. Necessity. Let $f(X)$ is a partially-defined threshold function, and $W = (w_0, w_1, ..., w_n)$ is its weighting vector. Let

$$F(\alpha) = \begin{cases} W(\alpha), & if\ \alpha \in U_1 \\ |W(\alpha)|e^{i\varphi(\alpha)}, & if\ \alpha \in U \end{cases}$$

where $\varphi(\alpha) = \arg(W(\alpha)) - \arg(f(\alpha))$. It is easy to convince that the conditions (2.4.23)-(2.4.24) are true for the function $F(X)$, which is defined in such a way.

Sufficiency. Let conditions of the theorem are true. Evidently, that the equality $P(F(\alpha)f(\alpha)) = f_1(\alpha)$ is true $\forall \alpha \in U_1$, and $f_1(\alpha) = f(\alpha) = P(W(\alpha))\ \forall \alpha \in U$. Theorem is proven.

Theorem 2.4.6. The partially defined function $f(X)$ of the k-valued logic $(k > 2)$ is a threshold function with a weighting vector $W = (w_0, w_1, ..., w_n)$, if and only if real-valued functions $A(X)$ and $B(X)$ defined on $\widetilde{G}$ exist such that

$$[A(X) + \varepsilon B(X)]f_1(X) = W(X), \tag{2.4.25}$$
$$\forall \alpha \in U\ A(X) > 0,\ B(X) \geq 0. \tag{2.4.26}$$

The proof of this theorem followes from the Theorem 2.2.4 and Corollary 2.2.1.

Theorem 2.4.7. The partially-defined function $f(X)$ of k-valued logic $(k > 2)$ is a threshold function with weighting vector $W = (w_0, w_1, ..., w_n)$ if and only if real-valued functions $A(X)$ and $B(X)$ defined on $\widetilde{G}$ exist such that

$$\left([A(X) + \varepsilon B(X)]f_1(X), \chi_j\right) = 0, \tag{2.4.27}$$

if $\chi_j \neq x_l$ for all the $l = 0, 1, 2, ..., n$, and

$$\forall \alpha \in U\ A(X) > 0, B(X) \geq 0. \tag{2.4.28}$$

If the conditions of the theorem are true then the weighting vector for the function $f(X)$ may be obtained in the following way:

$$w_i = \frac{1}{k^n}\left([A(X) + \varepsilon B(X)]f_1(X), X_i\right)\ i = 0, 1, ..., n. \tag{2.4.29}$$

The proof of this theorem follows from the Theorem 2.4.2.

It should be noted that the conditions of the Theorems 2.4.5 - 2.4.7 are not sufficient conditions for the thresholding property of the partially defined k-valued function $f_1(X)$, because the values of the function $F(X) = A(X) + \varepsilon B(X)$ are assumed as arbitrary on every $\alpha \in U_1$. It is also possible to change the condition $\forall \alpha \in U_1\ f_1(\alpha) = 1$ by $\forall \alpha \in U_1\ f_1(\alpha) > 0$.

So the synthesis of the MVN, which implements a mapping described by a partially-defined k-valued logic function $f(X)$, is reduced to search for such a solution $y = (y_1^0, \ldots, y_t^0, x_1^0, \ldots, x_m^0)$ of the system of linear algebraic homogeneous equations with complex-valued coefficients

$$a_{11}y_1 + a_{12}y_2 + \ldots + a_{1t}y_t + b_{11}x_1 + b_{12}x_2 + \ldots + b_{1m}x_m = 0,$$

$$\ldots\ldots\ldots\ldots\ldots\ldots\ldots\ldots\ldots\ldots\ldots\ldots\ldots\ldots\ldots\ldots\ldots\ldots\ldots\ldots \quad (2.4.30)$$

$$a_{r1}y_1 + a_{r2}y_2 + \ldots + a_{rt}y_t + b_{r1}x_1 + b_{r2}x_2 + \ldots + b_{rm}x_m = 0$$

that the following condition has to be true:

$$0 \le \arg(y_j^0) < \frac{2\pi}{k}, \quad j = 1, 2, \ldots, t. \quad (2.4.31)$$

Let us consider one of the possible approaches to solution of the problem. The system (2.4.30) may be transformed to the following:

$$b_{11}x_1 + b_{12}x_2 + \ldots + b_{1m}x_m = -\sum_{j=1}^{t} a_{1j}y_j,$$

$$\ldots\ldots\ldots\ldots\ldots\ldots\ldots\ldots\ldots\ldots\ldots\ldots\ldots\ldots \quad (2.4.32)$$

$$b_{r1}x_1 + b_{r2}x_2 + \ldots + b_{rm}x_m = -\sum_{j=1}^{t} a_{rj}y_j.$$

Let

$$B = \begin{pmatrix} b_{11} & b_{12} & \cdots & b_{1m} \\ b_{21} & b_{22} & \cdots & b_{2m} \\ \cdots & \cdots & \cdots & \cdots \\ b_{r1} & b_{r2} & \cdots & b_{rm} \end{pmatrix},$$

and R is a rank of the matrix B. One of the two alternatives is possible:

1. $R = r \le m$. The system (2.4.32) in such a case is reduced by sequential elimination of the unknowns to the following system (with a suitable renumbering of the unknowns):

$$b_{11}'x_1 + b_{12}'x_2 + \ldots + b_{1r}'x_r \ldots + b_{1m}'x_m = -\sum_{j=1}^{t} a_{1j}'y_j,$$

$$b_{22}'x_2 + \ldots + b_{2r}'x_r \ldots + b_{2m}'x_m = -\sum_{j=1}^{t} a_{2j}'y_j, \quad (2.4.33)$$

$$\ldots\ldots\ldots\ldots\ldots\ldots\ldots\ldots\ldots\ldots\ldots\ldots\ldots\ldots$$

$$b_{rr}'x_r \ldots + b_{rm}'x_m = -\sum_{j=1}^{t} a_{rj}'y_j,$$

where $b'_{ii} \neq 0$, $i = 1, 2, ..., r$. The values $y^0_1, ..., y^0_r$ of the solution should be chosen as arbitrary numbers satisfying the condition (2.4.31) (for example, the most easy solution is $y^0_1 = ... = y^0_r = 1$). For $r = m$ solution is trivial. For $r < m$ we can find the other components of our solution:

$$x^0_{r+1} = x^0_{r+2} = ... = x^0_m,$$

$$x^0_r = \frac{1}{b'_{rr}} \sum_{j=1}^{t} a'_{rj} y^0_j, \quad ..., \quad x^0_1 = \frac{1}{b'_{11}} \left(\sum_{j=1}^{t} a'_{1j} y^0_j - \sum_{j=2}^{t} b'_{1j} x^0_j \right).$$

2. $R < r$. The system (2.4.32) will be transformed in such a case to the following system:

$$b'_{11}x_1 + b'_{12}x_2 + ... + b'_{1R}x_R ... + b'_{1m}x_m = -\sum_{j=1}^{t} a'_{1j} y_j,$$

$$b'_{22}x_2 + ... + b'_{2R}x_R ... + b'_{2m}x_m = -\sum_{j=1}^{t} a'_{2j} y_j,$$

....

$$b'_{RR}x_R ... + b'_{Rm}x_m = -\sum_{j=1}^{t} a'_{Rj} y_j, \qquad (2.4.34)$$

$$0 = \sum_{j=1}^{t} a'_{(R+1)j} y_j,$$

....

$$0 = \sum_{j=1}^{t} a'_{rj} y_j.$$

Evidently, the system (2.4.34) has a solution $y = (y^0_1, ..., y^0_t, x^0_1, ..., x^0_m)$, which satisfies the condition (2.4.31) if and only if the system

$$0 = \sum_{j=1}^{t} a'_{(R+1)j} y_j,$$

.... (2.4.35)

$$0 = \sum_{j=1}^{t} a'_{rj} y_j$$

has a solution $y = (y^0_1, ..., y^0_t)$, which satisfies the same condition. The Theorem 2.4.3 shows the conditions of existing of such a solution of the system (2.4.35), which satisfies the condition (2.4.31). Let $y = (y^0_1, y^0_2, ..., y^0_t)$ is one of such solutions. Then all the other components of the solution $y = (y_1^0, ..., y_t^0, x_1^0, ..., x_m^0)$ of the system (2.4.30) may be found from the following system:

$$b'_{11}x_1 + b'_{12}x_2 + ... + b'_{1m}x_m = \sum_{j=1}^{t} a'_{1j} y_j^0,$$

$$b'_{22}x_2 + ... + b'_{2m}x_m = \sum_{j=1}^{t} a'_{2j} y_j^0,$$

(2.4.36)

....

$$b'_{RR}x_R + ... + b'_{Rm}x_m = \sum_{j=1}^{t} a'_{Rj} y_j^0$$

in a similar to the previous case way.

Example 2.4.4. Let $k=4$, $n=1$, $f(1)=1$, $f(-1)=i$, and function is not defined for $x = i$, $x = -i$. Let us consider the following function $f_1(1) = f_1(i) = f_1(-i) = 1; f_1(-1) = i$. According to the Theorem 2.4.5 we obtain the following:

$$(F(X)f(X), X^2) = F_0 - F_1 + iF_2 - F_3,$$

$$(F(X)f(X), X^3) = F_0 + iF_1 - iF_2 - iF_3.$$

We have to find a solution of the system

$$F_0 - F_1 + iF_2 - F_3 = 0,$$

$$F_0 + iF_1 - iF_2 - iF_3 = 0$$

such that $0 \le \arg(F_0) < \pi/2$; $0 \le \arg(F_2) < \pi/2$ will be true. After the transformations we obtain the following:

$$F_1 + F_3 + F_0 + iF_2,$$

$$iF_1 - iF_3 = F_0 + iF_2;$$

$$F_1 + F_3 + F_0 + iF_2,$$

$$2iF_1 + 0 \cdot F_3 = (i-1)F_0 + (i-1)F_2.$$

Let $F_0 = F_2 = 1$. Then $F_1 = \dfrac{2(i-1)}{2i} = 1 + i$; $F_3 = 1 + i - (i+1) = 0$. It means that $f(X)$ is threshold function, and we can find its weighting vector:

$$w_0 = \frac{1}{4}(F(X)f_1(X), \chi_0) = \frac{1}{4}(2 + 2i) = \frac{1+i}{2},$$

$$w_1 = \frac{1}{4}(F(X)f_1(X), X) = \frac{1}{4}(2 - 2i) = \frac{1-i}{2};$$

$$W = \left(\frac{1+i}{2}, \frac{1-i}{2}\right); \ W' = 2W = (1+i, 1-i).$$

5.   CONCLUSIONS

The theory of multiple-valued (k-valued) threshold functions over the field of the complex numbers based on group-theoretical approach has been developed in this Chapter. On the base of this theory the multi-valued neuron has been introduced as a neural element, which performs an input/output mapping described by k-valued threshold function. The following points are the key aspects for the presented approach: 1) values of the k-valued logic and the inputs and output of MVN are k^{th} roots of unity; 2) the activation function of the MVN is a function of the argument of the weighted sum. The features of multi-valued threshold functions are studied, and a geometrical (topological) interpretation of a mapping implemented by MVN has been considered. It was shown that a problem of the MVN synthesis might be reduced to the solution of a linear programming problem. The learning algorithm for MVN with two linear learning rules will be considered below in the Chapter 4, and different applications of MVN-based networks to solving of applied problems will be considered in the Chapters 5-6.

Chapter 3

P-Realizable Boolean Functions and Universal Binary Neurons

The universal binary neurons are a central point of this Chapter. The mathematical models of the universal binary neuron over the field of the complex numbers, the residue class ring and the finite field are considered. A notion of the *P*-realizable Boolean function, which may be considered as a generalization of a notion of the threshold Boolean function, is introduced. It is shown that the implementation with a single neuron the input/output mapping described by non-threshold Boolean functions is possible, if weights are complex, and activation function of the neuron is a function of the argument of the weighted sum (similar to multi-valued neuron). It is also possible to define an activation function on the residue class ring and the finite field in such a way that the implementation of non-threshold Boolean functions will be possible on the single neuron with weights from these sets. *P*-realization of multiple-valued function over finite algebras and residue class ring in particular is also considered. The general features of *P*-realizable Boolean functions are considered, also as the synthesis of universal binary neuron.

1. GENERALIZATIONS OF THE THRESHOLD ELEMENT OVER THE FIELD OF THE COMPLEX NUMBERS. NOTION OF *P*-REALIZABLE BOOLEAN FUNCTION

Intensive development of threshold logic was natural in the 1970-ies, because exactly within this field the features of threshold (linear-separable) Boolean functions (main objects implemented by classical perceptrons) have been deeply studied. The fundamental properties of threshold logic have been derived e.g. in [Dertouzos (1965), Muroga(1971)]. Principal limitations of perceptrons were discovered in [Minsky & Papert (1969)]. After this publication almost all the efforts in the field have been directed on hardware and VLSI implementation of the threshold logic gates and development of the different networks for implementation of the non-threshold Boolean functions.

Let us return to the one of the equivalent definitions of the Boolean threshold function (Definition 1.1.2). The Boolean function $f(x_1,...,x_n)$ is *threshold (linear-separable)*, if it is possible to find a real-valued weighting vector $W = (w_0, w_1,..., w_n)$ such that the equation

$$f(x_1,...x_n) = sign(w_0 + w_1 x_1 +...+ w_n x_n) \qquad (3.1.1)$$

will be true for all the values of the variables x from the domain of the function f. The number of threshold functions for $n \geq 3$ is limited (see Section 1.1 for some details). Is it possible to produce another representation of Boolean function with $n+1$ weights, which makes possibile to implement at least some of non-threshold functions or even all the functions of n variables?

What is the *sign* function? It should be considered as a binary predicate. It is possible to consider other predicates, and to expand the set of threshold functions in such a way. May be one of the first ideas in this direction has been proposed in [Askerov (1965)]. It was proposed to consider instead of threshold activation function the following piece-wise "multi-threshold" function:

$T(z) =$

$$= \begin{cases} 1, & \textit{if } z \in [a_1, a_2[\textit{ or } z \in [a_3, a_4[\textit{ or } ... \textit{ or } z \in [a_{k-2}, a_{k-1}[\textit{ or } z \in [a_k, \infty[\\ -1, & \textit{if } z \in]-\infty, a_1[\; z \in [a_2, a_3[\textit{ or } z \in [a_4, a_5[\textit{ or } ... \textit{ or } z \in [a_{k-1}, a_k[. \end{cases} \qquad (3.1.2)$$

The equation (3.1.2) is illustrated in Fig. 3.1.1.

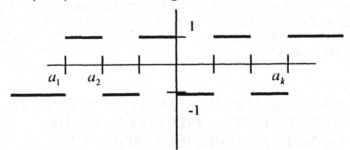

Figure 3.1.1. "Multi-threshold" activation function

Despite the fact that it was possible to implement by mean of this function (3.1.2) many of the non-threshold functions, no learning algorithm has been developed for choice also as choice of the intervals $[a_i, a_{i+1}[$.

An attempt to extend a set of the threshold-like represented functions by consideration of the complex-valued weights has been done in [Aizenberg N. et al. (1973a)] and [Aizenberg N. & Ivaskiv (1977)]. The next statement follows directly from the Definition 1.1.2 of threshold function.

Theorem 3.1.1. The Boolean function $f(x_1,...,x_n)$ is a threshold function if and only if a real-valued weighting vector $W = (w_0, w_1,..., w_n)$ exists such that

$$\forall \alpha \in \tilde{E}_2^n \quad \arg(f(\alpha)) = \arg(W(\alpha)).$$

Let define the following functions-predicates (see also Fig. 3.1.2):

$$Rsign(z) = \begin{cases} 1, & \textit{if } \arg(z) = 0 \\ -1, & \textit{if } \arg(z) = \pi \end{cases}, \qquad (3.1.3)$$

$$Csign(z) = \begin{cases} 1, & \textit{if } 0 \le \arg(z) < \pi \\ -1, & \textit{if } \pi \le \arg(z) < 2\pi \end{cases}. \qquad (3.1.4)$$

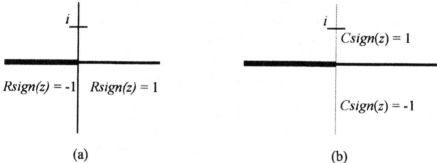

(a)   (b)

Figure 3.1.2. (a) - definition of the function *Rsign*; (b) - definition of the function *Csign*

Definition 3.1.1. The Boolean function $f(x_1,...,x_n)$ is *real*-threshold, if a complex-valued weighting vector $W = (w_0, w_1,..., w_n)$ exists such that the weighted sum $W(X)$ is always real-valued, and $f(x_1,...,x_n) = Rsign(W(X))$ for the whole domain of the function $f(x_1,...,x_n)$. The Boolean function $f(x_1,...,x_n)$ is *complex*-threshold, if a complex-valued weighting vector $W = (w_0, w_1,..., w_n)$ exists such that $f(x_1,...,x_n) = Csign(W(X))$ for the whole domain of the function $f(x_1,...,x_n)$.

It should be noted that definition of the complex-threshold Boolean function coincides with the definition of the *k*-valued threshold function (see Definition 2.2.1) for *k*=2.

Let T be a set of the Boolean threshold functions, T_R is a set of the Boolean real-threshold functions and T_c is a set of the Boolean complex-threshold functions.

It is absolutely clear that the Boolean function is threshold if and only if a real-valued weighting vector $W = (w_0, w_1, \ldots, w_n)$ exists such that $f(x_1, \ldots, x_n) = Rsign(W(X))$. From this fact, and from the Definition 3.1.1 the following theorem holds.

Theorem 3.1.2. $T \subseteq T_R \subseteq T_C$.

But the following theorem is more important.

Theorem 3.1.3. $T_C \subseteq T$.

Proof. Let α pass through the set $\tilde{E}_2^n$ and $f(X) \in T_C$. It means that such a weighting vector $W = (w_0, w_1, \ldots, w_n)$ exists that $f(x_1, \ldots, x_n) = Csign(W(X))$. Let V be a set of all the values $W(\alpha)$, $\alpha \in \tilde{E}_2^n$. Evidently, cardinality of the set V is not greater than 2^n: $|V| \le 2^n$.

We have to consider the following cases: 1) There are no truly real numbers in V; 2) It is at least single truly real number in V.

In the first case $Im(W(\alpha)) \ne 0$ for any $\alpha \in \tilde{E}_2^n$ and therefore

$$f(X) = Csign(W(X)) = Csign(Im(W(X)) =$$
$$= Csign[Im(w_0 + w_1 x_1 + \ldots + w_n x_n)] =$$
$$= Csign[Im\, w_0 + x_1 Im\, w_1 + \ldots + x_n Im\, w_n]$$

$(Im\, w_0 + x_1 Im\, w_1 + \ldots + x_n Im\, w_n) \in R$, and it follows from the equalities (3.1.3) and (3.1.4) that

$$Csign[Im\, w_0 + x_1 Im\, w_1 + \ldots + x_n Im\, w_n] =$$
$$= Rsign[Im\, w_0 + x_1 Im\, w_1 + \ldots + x_n Im\, w_n]$$

This means that $f(X) = Rsign[Im\, w_0 + x_1 Im\, w_1 + \ldots + x_n Im\, w_n]$ and taking into account that $Im\, w_j \in R$, $j = 0, 1, \ldots, n$ we can conclude that $f(X) \in T_R$, and $f(X) \in T$ simultaneously.

To consider the second case we will replace the set V with the set

$$V' = \left\{ W'(\alpha) | W'(\alpha) = W(\alpha) e^{i\varphi}, \ \alpha \in \tilde{E}_2^n \right\},$$

where $\varphi = \dfrac{1}{2} \min \begin{Bmatrix} \pi - \max_{z \in V'}(\arg(z)), & 2\pi - \max_{z \in V}(\arg(z)), \\ 0 \le \arg(z) < \pi, & \pi \le \arg(z) < 2\pi \end{Bmatrix}$. It is evident

that $0 < \varphi \le \pi/2$. The set V' consists of all the elements of the set V rotated over the same angle φ with the aim to satisfy two conditions: 1) to ensure absence of at least single truly real number within set V'; 2) To ensure that the following condition will be true: $\forall \alpha \in \widetilde{E}_2^n$ $Csign(W(\alpha)) = Csign(W'(\alpha))$. Let $w_j' = w_j e^{ij}$, $j = 0, 1, ..., n$. It ensures that $Im(W'(\alpha)) \neq 0$, and moreover, according to the choice of φ $Csign(w_0 + w_1 x_1 + ... + w_n x_n) = Csign(w_0' + w_1' x_1 + ... + w_n' x_n)$. This means that the case 2 is reduced to the case 1, for which theorem has been already proven.

The following important corollary follows from the Theorems 3.1.2 and 3.1.3.

Corollary 3.1.1. The sets T of Boolean threshold functions, T_R of Boolean real-threshold functions and T_C of Boolean complex-threshold functions coincide with each other: $T = T_R = T_C$.

So the Definition 3.1.1 expands the possible values of weights of the Boolean threshold functions to the field of the complex numbers, but does not expand a number of Boolean threshold functions.

At first sight the introduction of the notion of complex-threshold and real-threshold Boolean functions does not create new possibilities. On the other hand it shows that number of linear-separable Boolean functions does not increase with moving to the field of the complex numbers, but with preservation of threshold type of the activation function.

The following generalization of the Definitions 1.1.2 and 3.1.1 has been proposed in [Aizenberg I. (1985)].

Definition 3.1.2. The Boolean function $f(x_1, ..., x_n)$ is called *P-realizable* over the algebra A, if it is possible to define a predicate P_B on the algebra A and to find a weighting vector $W = (w_0, w_1, ..., w_n)$, $w_j \in A$, $j=0, 1, ..., n$ such that the equality

$$f(x_1, ... x_n) = P_B(w_0 + w_1 x_1 + ... + w_n x_n) \qquad (3.1.5)$$

holds for all the values of the variables x in the domain of the function f.

Evidently, if $A = R$ and P_B is the function *sign*, we obtain from the Definition 3.1.2 the Definition 1.1.2 of Boolean threshold function. If $A = C$ and P_B is the function *Csign*, or *Rsign* we obtain from the Definition 3.1.2 the Definition 3.1.1 of Boolean complex-threshold, or real-threshold function, respectively. At the same time it is possible to consider A not only as the field of the complex or real numbers, but for example, as residue class ring Z_m or finite field GF. It is also possible to define other predicates P_B over the field of the complex numbers and to consider some predicates over Z_m and GF with the aim to increase a number of Boolean functions which may be presented like (3.1.5) in comparison with a number of threshold

Boolean functions. Solving this problem it is possible to develop a neural element with the activation function P_B.

Let $A = C$, so the field of the complex numbers is our basic set. To fill the Definition 3.1.2 by concrete contents let us define the function P_B in the following way (we already described this approach prospectively in the Section 1.2):

$$P_B(z) = \begin{cases} 1, \text{if } 0 \le \arg(z) < \pi/2 \text{ or } \pi \le \arg(z) < 3\pi/2 \\ -1, \text{if } \pi/2 \le \arg(z) < \pi \text{ or } 3\pi/2 \le \arg(z) < 2\pi \end{cases}. \quad (3.1.6)$$

The equality (3.1.6) is illustrated in Fig. 3.1.3.

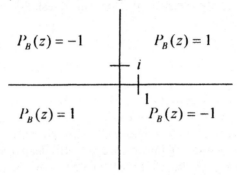

$P_B(z) = -1$ $P_B(z) = 1$

i

1

$P_B(z) = 1$ $P_B(z) = -1$

Figure 3.1.3 Illustration for the equality (3.1.6)

So the function (3.1.6) separates the complex plane onto four equal sectors(quadrants): in two of them $P_B = 1$, in other two of them $P_B = -1$. As it was mentioned above (Section 1.2) the XOR problem for two inputs may be solved on a single neuron with activation function (3.1.6) for example, by weighting vector $(0, 1, i)$ (see Table 1.2.1). It means that traditional opinion that XOR problem can not be solved on a single neuron (the corresponding example may be found almost in any book devoted to the neural networks, we address reader e.g., to the recent publications [Haykin (1994)], [Hassoun (1995)]) is false.

Let us consider some features, which follow from the Definition 3.1.2 and the equality (3.1.6). We will consider here for simplicity the Boolean functions of two variables and then we will generalize the corresponding features to the functions of n variables.

Let us suppose that Boolean function $f(x_1, x_2)$ is P-realizable (the function P_B is defined by the equality (3.1.6)). It means that if $f(\alpha_1, \alpha_2) = f(\alpha) = 1$, then a number $r_\alpha \in C$ exists such that

$$w_0 + w_1\alpha_1 + w_2\alpha_2 = W(\alpha) = r_\alpha f(\alpha_1, \alpha_2) = r_\alpha f(\alpha), \quad (3.1.7)$$

and evidently

$$(0 \leq \arg(r_\alpha) < \pi/2) \vee (\pi \leq \arg(r_\alpha) < 3\pi/2). \qquad (3.1.8)$$

If $f(\alpha_1, \alpha_2) = f(\alpha) = -1$, then the equality (3.1.7) is true, but for r_α we have the following

$$(\pi/2 \leq \arg(r_\alpha) < \pi) \vee (3\pi/2 \leq \arg(r_\alpha) < 2\pi). \qquad (3.1.9)$$

It is clear from (3.1.7) - (3.1.9) that if $f(\alpha) = 1$, then

$$r_\alpha = W(\alpha); \quad |W(\alpha)| = |r_\alpha|, \qquad (3.1.10)$$

and if $f(\alpha) = -1$ then

$$r_\alpha = -W(\alpha); \quad |W(\alpha)| = |r_\alpha|. \qquad (3.1.11)$$

Theorem 3.1.4. The Boolean function $f(x_1, x_2)$ is P-realizable with the weighting vector $W = (w_0, w_1, w_2)$, if for any values $(\alpha_1, \alpha_2) = (\alpha)$ from the domain of function f such a number $r_\alpha \in C$ exists that for $f(\alpha) = 1$ (3.1.7), (3.1.8), (3.1.9) are true, and for $f(\alpha) = -1$ (3.1.7), (3.1.9), (3.1.11) are true.

Necessity has been proven with considerations, by which the formulas (3.1.7)-(3.1.11) have been obtained. Sufficiency may be proven by inverse considerations.

It should be noted that vectors $X_1, ..., X_n$ of the Boolean variables $x_1, ..., x_n$, in alphabet $\{1, -1\}$ and vector $X_0 = (1, ..., 1)$ are the characters of the group E_2^n with the numbers 0, 2^{n-j-1}, $j=1$, ..., n, and simultaneously they completely coincide with the Rademacher functions [Rademacher (1922)]. To simplify a numbering of the characters we will use the decimal representation of numbers that are formed by binary vectors, which are elements of the set E_2^n. We have to remind that features of the group characters have been considered in the Section 2.1, and an example for $n=2$ is presented by equation (2.1.27). To consider the features of the P-realizable Boolean functions in details we will use the features of the characters.

Let us renumber the characters of the group E_2^n similar to the renumbering of the characters of the group E_k^n in the Section 2.4: $\chi_0 = X_0, \chi_1 = X_1, ..., \chi_n = X_n, \chi_{n+1}, ..., \chi_{2^n-1}$, so $\chi_{n+1}, ..., \chi_{2^n-1}$ are all the characters, which are different from the Rademacher functions. Let us denote the vector of the values of Boolean function $f(x_1, ..., x_n)$ as $f = (f_0, f_1, ..., f_{2^n-1})$.

Theorem 3.1.5. The Boolean function $f(x_1, x_2)$ is P-realizable if and only if a complex-valued vector $r = (r_0, r_1, r_2, r_3)$, $r_j \neq 0$, $j = 0, 1, 2, 3$ exists such that the equality

$$(r \circ f, \chi_3) = 0, \qquad (3.1.12)$$

where $\circ$ denotes the component by component multiplication of the vectors, is true, and all the components of the vector r r_j, $j = 0, 1, 2, 3$ satisfy the conditions (3.1.7), (3.1.8), (3.1.10), if $f_j = 1$ or the conditions (3.1.7), (3.1.9), (3.1.11) if $f_j = -1$.

Proof. Necessity. Let Boolean function $f(x_1, x_2)$ be P-realizable with the weighting vector $W = (w_0, w_1, w_2)$. According to the Theorem 3.1.4 a set of $r_\alpha \in C$, $\alpha = 0, 1, 2, 3$ exists that $\forall \alpha \ W(\alpha) = r_\alpha f(\alpha)$, and the conditions (3.1.7) - (3.1.11) are true for r_α, $\alpha = 0, 1, 2, 3$. Since the characters establish a basis, the following decomposition takes place, and it is unique:

$$r \circ f = \gamma_0 X_0 + \gamma_1 X_1 + \gamma_2 X_2 + \gamma_3 \chi_3 \qquad (3.1.13)$$

(the only character χ_3 of the group E_2^2 does not coincide with any Rademacher function). But since the decomposition (3.1.13) is unique, from it and from the equality (3.1.7) we can to conclude that $\gamma_0 = w_0$; $\gamma_1 = w_1$; $\gamma_2 = w_2$; $\gamma_3 = 0$. It means that the vector r satisfies the conditions of the theorem because (3.1.12) is true if $\gamma_3 = 0$.

Sufficiency. Let the conditions of the theorem be true. The decomposition (3.1.13) is unique, and according to (3.1.12) $\gamma_3 = 0$. It means that $\gamma_0 = w_0$; $\gamma_1 = w_1$; $\gamma_2 = w_2$, and according to the Theorem 3.1.4 $W = (w_0, w_1, w_2)$ is the weighting vector of the function $f(x_1, x_2)$, which is P-realizable.

Theorem 3.1.6. If the Boolean function $f(x_1, x_2)$ is P-realizable then the components of its weighting vector may be obtained in the following way:

$$w_j = (r^* \circ f, X_j), \ j = 0, 1, 2, \qquad (3.1.14)$$

where $r^* = (r_0^*, r_1^*, r_2^*, r_3^*)$ is a solution of the equation (3.1.12) with respect to r.

Proof. The function $f(x_1, x_2)$ is P-realizable, and the conditions of the Theorem 3.1.5 are true for it. The equation (3.1.13) for the function $f(x_1, x_2)$ may be transformed to the $w_0 X_0 + w_1 X_1 + w_2 X_2 = r^* \circ f$. Let us evaluate a scalar product of the last equality with X_0, X_1, X_2 consecutively: $w_j (X_j, X_j) = (r^* \circ f, X_j)$, $j = 0, 1, 2$. Taking into account that the characters are orthogonal, we obtain

$$w_j = \frac{1}{4}(r^* \circ f, X_j), \; j = 0, 1, 2.$$ Finally, taking into account that

$\arg ar_\alpha = \arg r_\alpha$, $a > 0$, $r_\alpha \in C$, we obtain the equalities (3.1.14) for components of the vector W.

Let us apply the Theorems 3.1.5 - 3.1.6 to obtain a weighting vector for an interesting example.

Example 3.1.1. Let us take the XOR, which is the favorite example of non-linearly separable problem. $XOR(x_1, x_2) = X_1 \circ X_2 = (1, -1, -1, 1)$. The equation (3.1.12) for this function will be the following: $\big((r_0, r_1, r_2, r_3),(1, -1, -1, 1) \circ (1, -1, -1, 1)\big) = 0$ and evaluating the scalar product: $r_0 + r_1 + r_2 + r_3 = 0$. According to the conditions (3.1.7)-(3.1.11) r_0 and r_3 have to be within 1st or 3rd quadrant of the complex plane, r_1 and r_2 have to be within 2nd or 4th ones. One of the possible and acceptable solutions is for example the following: $r_0^* = 1 + i$; $r_1^* = 1 - i$; $r_2^* = -1 + i$; $r_3^* = -1 - i$. According to the equalities (3.1.14) $w_0 = 0$, $w_1 = 4i$, $w_2 = 4$, or dividing all the w by 4 we obtain the more "beautiful" components of the weighting vector $w_0 = 0$, $w_1 = i$, $w_2 = 1$. Table 3.1.1 illustrates that the weighting vector $(0, i, 1)$ solves the XOR problem for two inputs. One may compare the Table 1.3.1 with the Table 1.2.1, which illustrates solution of the XOR problem for another weighting vector $(0, 1, i)$, but with the same activation function (3.1.6). We will show in the Chapter 4, that the learning algorithms can also be found for the XOR and other non-threshold Boolean functions.

Table 3.1.1. Solution of the XOR problem

x_1	x_2	$z = w_0 + w_1 x_1 + w_2 x_2$	$P_B(z)$	XOR
1	1	$1 + i$	1	1
1	-1	$-1 + i$	-1	-1
-1	1	$1 - i$	-1	-1
-1	-1	$-1 - i$	1	1

It is easy to confirm by a simple computing experiment that the equation (3.1.12) has an infinite number of acceptable solutions for any of the sixteen Boolean functions of two variables. This means that all of them are *P*-realizable with predicate (3.1.6). We will return below to the consideration of features of the predicate likes (3.1.6). To finish this Section we present the Table 3.1.2, which contains examples of the weighting vectors for all Boolean functions of two variables. All the vectors are obtained in the same way as for the XOR function (i.e. using Theorems 3.1.4-3.1.6). The Boolean functions are numbered by decimal numbers, which are decimal representations of the binary numbers created by concatenation of all the values of function in alphabet $\{0, 1\}$ ($f_0 = (0,0,0,0)$; $f_1 = (0,0,0,1)$, etc.).

Table 3.1.2. Weighting vectors of the Boolean functions of two variables relatively the predicate (3.1.6).

Func #	0	1	2	3	4	5	6	7	8	9	10	11	12	13	14	15
w_0	1	1	i	i	1	i	0	1	i	0	-1	i	-1	-1	i	i
w_1	0	i	1	1	$-i$	0	1	1	i	i	0	1	i	i	-1	0
w_2	0	i	i	0	i	1	i	i	-1	-1	i	-1	0	-1	-1	0

The following existence theorem shows that problem of P-realization of the Boolean functions of n variables over the field of the complex numbers may be solved for any n, but with predicate not like (3.1.6).

Theorem 3.1.7. For any value of n it is possible to define the predicate P_B in such a way that it ensures a P-realization of all the Boolean functions of n variables.

Proof. For $n=2$ the theorem has been proven by the above considerations. The proof for $n \geq 3$ will be done geometrically.

Let partition the complex plane on $2^{2^n} = N$ equal sectors with the angle value $2\pi / N$. Let us consider the binary sequences, which have the following property: all the relevant subsequences occur among the subsequences of length 2^n. The length of such a sequence is at least 2^{2^n}, and as we know they exist for any n [Peterson & Weldon (1972)]. The simplest of such sequences, or a complete code ring for code sequences of a length 2^n, is the complete code ring obtained from the maximum-length sequences, or M-sequences. The M-sequence is a binary recursive sequence of a special form [Peterson & Weldon (1972)]. Let us consider the M-sequence $\{m_i\}$ with the period $2^{2^n} - 1$ with the first $2^n - 1$ zero-valued terms. Create the sequence $\{a_i\}$ according to the rule:

$$a_i = m_i, \ i = 1, \ ..., \ 2^{2^n} - 1;$$
$$a_0 = 0; \ a_{2^{2^n}-1+i} = a_i. \tag{3.1.15}$$

Evidently, the sequence $\{a_i\}$ contains the values of all Boolean functions of n variables. Let S_i; $i = 0, 1, ..., N-1$ be the ith sector on the complex plain. Let us define the predicate $P_B = P$ on the field of the complex numbers in the following way:

$$P(u) = a_i, \quad if \ u \in S_i; \ i = 0, 1, \ldots, N-1, \qquad 3.1.16a)$$

or, which is the same,

$$P(u) = a_i, \quad if \ \frac{2\pi}{N} \le \arg(u) < \frac{2\pi}{N}(i+1); \ i = 0, 1, \ldots, N-1. \qquad (3.1.16b)$$

Let the numbering of the sectors starts from the real axis:

$$u \in S_0 \Rightarrow 0 \le \arg(u) < 2\pi / n, \text{ etc.}$$

Let consider case $n=3$, which is illustrated in the Fig. 3.1.4.

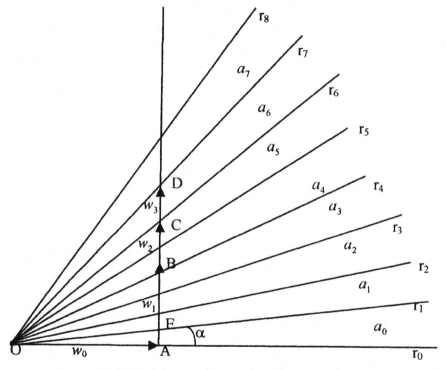

Figure 3.1.4. Illustration to the proof of the Theorem 3.1.7

Define the boundaries of the sector S_i as the rays r_i and r_{i+1}. Let the sequence $\{a_i\}$ contains the values of some function $f_i(y_1, \ldots, y_n)$ beginning from some term. Choose the nonzero weight w_0 arbitrarily on the real axis (therefore it will be truly real number). Let $w_0 = |OA|$. Construct the line perpendicular to the real axis in the point A. Let B be the intersection point of this perpendicular with the ray r_4, C- with the ray r_6, D - with the ray r_7, E - with the ray r_1. Let choose the weights w_1, w_2, w_3 truly imaginary, but such that the following correspondence will be true for their absolute values:

$$|w_1|=|AB|, \ |w_2|=|BC|, \ |w_3|=|CD|.$$

It should be noted that according to our numbering of the sectors our function f_i should be exactly the function f_0, but it will be shown that all our considerations will be true for any function. To ensure the P-realization of the function f_i it is necessary to ensure that (3.1.16) is true for any combination of the values of variables. Let $z(\alpha_1, \alpha_2, \alpha_3)$ be the value of the weighted sum corresponding to the values $(\alpha_1, \alpha_2, \alpha_3)$ of variables. It means that for P-realizability of the function f_i the following equalities should be true:

$$
\begin{array}{ll}
z(0, \ 0, \ 0) = w_0 \in S_0 & (a) \\
z(0, \ 0, \ 1) = (w_0 + w_3) \in S_1 & (b) \\
z(0, \ 1, \ 0) = (w_0 + w_2) \in S_2 & (c) \\
z(0, \ 1, \ 1) = (w_0 + w_2 + w_3) \in S_3 & (d) \\
z(1, \ 0, \ 0) = (w_0 + w_1) \in S_4 & (e) \\
z(1, \ 0, \ 1) = (w_0 + w_1 + w_3) \in S_5 & (f) \\
z(1, \ 1, \ 0) = (w_0 + w_1 + w_2) \in S_6 & (g) \\
z(1, \ 1, \ 1) = (w_0 + w_1 + w_2 + w_3) \in S_7 & (h).
\end{array}
\qquad (3.1.17)
$$

According to choice of the weights the equalities (3.1.17 a, e, g, h) are true, which is evident from the Fig. 3.1.4. Since the triangles OAE, OAB, OAC, OAD are rectangular then:

$$
\begin{array}{l}
|w_1|=|w_0| \cdot tg(4\alpha); \\
|w_2|=|w_0| \cdot [tg(6\alpha) - tg(4\alpha)]; \\
|w_3|=|w_0| \cdot [tg(7\alpha) - tg(6\alpha)]; \\
|w_2 + w_3|=|w_0| \cdot [tg(7\alpha) - tg(4\alpha)]; \\
|w_1 + w_3|=|w_0| \cdot [tg(4\alpha) + tg(7\alpha) - tg(6\alpha)]; \\
|AE|=|w_0| \cdot tg(\alpha),
\end{array}
\qquad (3.1.18)
$$

where $\alpha = 2\pi / 256 = \pi / 128$. To ensure that the equality (3.1.17b) will be true it is necessary to verify that the following inequality is true

$$tg(\alpha) \le tg(7\alpha) - tg(6\alpha) < tg(2\alpha).$$

This equality is true for the given value of angle α ($\alpha = 2\pi/256 = \pi/128$). From the same considerations the following inequalities are true:

$$tg(2\alpha) \le tg(6\alpha) - tg(4\alpha) < tg(3\alpha)$$
$$tg(3\alpha) \le tg(7\alpha) - tg(4\alpha) < tg(4\alpha)$$
$$tg(5\alpha) \le tg(4\alpha) + tg(7\alpha) - tg(6\alpha) < tg(6\alpha).$$

It means that the equalities (3.1.17 c, d, f) are also true and the function f_i (f_0) is P-realizable. Let us show that all other functions also are P-realizable. Let the sequence $\{a_i\}$ contain the values of the function f_j; $1 \le j < 255$ beginning from the term number k. It means that $P(S_k) = f_j(0, 0, 0)$. Let us choose the weight w_0 on the ray r_k arbitrarily (to illustrate this, it is enough to "rotate" the complex plane (Fig. 3.1.4) over some angle and to substitute respectively, $r_0, ..., r_8$ and $\alpha_0, ..., \alpha_7$ with $r_k, ..., r_{k+8}$ and $\alpha_k, ..., \alpha_{k+7}$). Let $|OA| = |w_0|$. Let us build a perpendicular to the ray r_k in the point A. The points of the intersection of such a perpendicular with the rays r_{k+4}, r_{k+6}, r_{k+7} are B, C and D.

The weight w_1 it is necessary to choose in such a way that $w_0 + w_1 \in S_{k+5}$. Let

$$w_1 = \vec{AB} = \vec{OB} + \vec{AO} = \vec{OB} - \vec{OA};$$

$$w_2 = \vec{OC} - \vec{OB};$$

$$w_3 = \vec{OD} - \vec{OC}.$$

Consideration of the equalities like (3.1.17) for the weighting sums leads us to the equalities that are equivalent to (3.1.18). These equalities are true, as has been just shown. Therefore an arbitrary function of three variables is P-realizable and, that is the same, all functions of three variables are P-realizable. For $n > 3$ all the considerations are analogous. Let the sequence $\{a_i\}$ contain the values of the function f_j; $1 \le j < 2^{2^n} - 1$ beginning from the term number k. It means that $P(S_k) = f_j(0, ..., 0)$. Let us choose the weight w_0 on the ray r_k arbitrary. Let $|OA| = |w_0|$, where point O corresponds to the complex number $(0, 0)$, and A is a point on the ray r_k. Let us construct again a line perpendicular to the ray r_k in the point A. The weight w_1 has to be chosen such that $w_1 = \vec{AA}_{k+2^{n-1}}$, where $A_{k+2^{n-1}}$ is a point of intersection of the perpendicular with the ray $r_{k+2^{n-1}}$. Let $A_{k+2^{n-1}+2^{n-2}}$ is a point of intersection of the perpendicular with the ray

$r_{k+2^{n-1}+2^{n-2}}$, etc., A_m is the point of intersection of the perpendicular with the ray r_m. The weight w_2 has to be chosen such that

$$w_2 = \overline{A_{k+2^{n-1}} \quad A_{k+2^{n-1}+2^{n-2}}} \quad , \text{ etc., the weight } w_n \text{ has to be chosen such that}$$

$$w_n = \overline{A_{k+2^{n-1}+\ldots+2} \quad A_{k+2^{n-1}+2^{n-2}+\ldots+2+1}} \quad . \text{ Evidently, considering the equalities for the}$$

weighting sums, we will obtain the equalities like (3.1.17), and then the equalities like (3.1.18). The value of angle α is going closer to zero with increasing of n, and we can conclude that equalities like (1.3.18) will true for any $n \geq 3$. Therefore an arbitrary function of n variables is P-realizable. It means that $(\forall n)$ all functions of n variables are P-realizable.

2.   *P*-REALIZABLE BOOLEAN FUNCTIONS OVER THE FIELD OF THE COMPLEX NUMBERS

We have just convinced that it is possible to define the activation function P_B as the alternating sequence (3.1.6) over the field of the complex numbers that all Boolean functions of two variables including not threshold function XOR and not-XOR will be P-realizable according to the Definition 3.1.2. Let us consider a general problem (mainly based on paper [Aizenberg I. (1991)] with some recent additions).

The equality (3.1.6), which defines a predicate implementing all the function of two variables, separates the complex plane in 2 x 2=4 equal sectors. Let us define the predicate P_B over the field of the complex numbers in the following way. Let k be a natural number. We will separate the complex plane on $m = 2k$ equal sectors and the predicate P_B will be defined by following equation:

$$P_B(z) = (-1)^j, \text{ if } 2\pi(j+1)/m > \arg(z) \geq 2\pi j/m, \quad m = 2k, \; k \in N. \quad (3.2.1)$$

where i is imaginary unity, m is some even positive integer, j is a non-negative number $0 \leq j < m$. So if we separate complex plane on m equal sectors, the function P_B is equal to 1 for the complex numbers in the even sectors 0, 2, 4, ..., m-2, and it is equal to -1 for the numbers in the odd sectors 1, 3, 5, ..., m-1. The equation (3.2.1) is illustrated in Fig. 3.2.1. One may compare a definition of the function P_B with a definition of the function P , which is an activation function of the multi-valued neuron (the equality (2.2.2), Fig. 2.2.1). Both functions separate the complex plane onto even number of equal sectors. The equality (3.2.1) does not define the value $P_B(0)$. Without loss of generality let us set $P_B(0) = 1$.

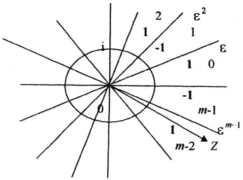

$$P_B(z) = 1$$

Figure 3.2.1. Definition of the function P_B (see the equation (3.2.1))

Definition 3.2.1. The Boolean function $f(x_1,...,x_n)$ is called *P-realizable* over the field of the complex numbers, if it is possible to find a complex-valued weighting vector $(w_0, w_1,..., w_n)$ such that the equation

$$f(x_1,...x_n) = P_B(w_0 + w_1x_1 + ... + w_nx_n) \qquad (3.2.2)$$

(where P_B is defined by the equality (3.2.1)) will be true for all values of the variables x from the domain of the function f.

Definition 3.2.2. The *Universal binary neural element* (*UBN*) over the field of the complex numbers is an element, which performs the equality (3.2.2) with the activation function (3.2.1) for a given input/output mapping described by a Boolean function of n variables.

Taking $k=1$ and therefore $m=2$ in (3.2.1) we obtain the equality (3.1.4), which defines the function *Csign*. A notion of the *P*-realizable Boolean function in such a case coincides with a notion of the complex-threshold Boolean function (see Definition 3.1.1), and universal binary neuron becomes an obvious threshold neuron. As it was shown above the activation function *Csign* does not increase functionality of a neuron in comparison with the threshold activation function. *The functionality of the UBN with the activation function (3.2.1) however will be always higher than the functionality of the neuron with the threshold activation function* for $k>1$ and $m>2$ in (3.2.1). For example, for $k=2$, $m=4$ UBN has complete functionality for $n=2$, and its functionality for $n>2$ is always much more higher than functionality of the threshold neuron.

Let us consider the character χ_k from the group of characters of the cyclic group $G = \{0, 1, ..., 2k-1\} = E_{2k}$. This group of an order $m=2k$ (k is chosen from (3.2.1)) is a group for the mod $2k$ addition. Evidently, $\chi_k = (1, -1, 1, -1, ..., 1, -1)$, or $\chi_k = (1)^g$, $g \in G$. Each of the $m=2k$ sectors on the complex plane, which are established by the function (3.2.1), may be numbered by the elements of the group G, or by values of the generating element of the group of characters $\chi_1 = (1, \varepsilon, \varepsilon^2, \varepsilon^3, ..., \varepsilon^{2k-1})$, where $\varepsilon = \exp(2\pi i / 2k)$ is a primitive $2k^{th}$ root of a unity (see Fig. 3.2.1). If $z \in C$ belongs to the jth sector on the complex plane (according to (3.2.1)), the function P_B may be defined in the following equivalent way (one may compare with (3.2.1)):

$$P_B(z) = \chi_k(j), \quad j = 0, 1, ..., 2k-1, \qquad (3.2.3)$$

that is the value of P_B on the jth sector is equal to the value of the character χ_k on the element $j \in G$.

Let us consider the mapping φ_j:

$$\forall g \in G \ \varphi_j\big(\chi_j(g)\big) = \chi_k(g), \quad j = 0, 1, ..., 2k-1, \qquad (3.2.4)$$

which establishes a correspondence between the characters χ_j and χ_k. Evidently, the inverse mapping, which establishes a correspondence between the characters χ_k and χ_j, also exists. It is easy to convince comparing the equalities (3.2.3) and (3.2.4) that the mapping φ_1: $\chi_1(g) \rightarrow \chi_k(g)$ defines the function P_B.

The following theorem generalizes the Theorem 3.1.4 for a function of two variables on a function of n variables.

Theorem 3.2.1. The Boolean function $f(x_1,...,x_n)$ is P-realizable over the field of the complex numbers if and only if for any set $\alpha = (\alpha_1, ..., \alpha_n)$ of the values of Boolean variables from the domain of the function f a number $r_\alpha \in C$ exists (or, such a 2^n-dimensional complex-valued vector $r = (r_{\alpha_0}, r_{\alpha_1}, ..., r_{\alpha_{2^n-1}})$ exists for fully defined Boolean function) such that the following equivalent equalities hold:

$$w_0 + w_1 X_1 + ... + w_n X_n = r \circ f \qquad (3.2.5a)$$

$$w_0 + w_1 \alpha_1 + ... + w_n \alpha_n = r_\alpha f(\alpha_1,...,\alpha_n). \qquad (3.2.5b)$$

The components of the vector r have to satisfy the following conditions: if $f(\alpha_1,...,\alpha_n) = 1$ r_α belongs to one of the sectors whose number belongs to the set $\varphi_{k-1}^{-1}(1)$, if $f(\alpha_1,...,\alpha_n) = -1$ r_α belongs to one of the sectors whose number belongs to the set $\varphi_{k+1}^{-1}(-1)$.

Proof. Necessity. $\varphi_{k-1}^{-1}(1)$ is a complete prototype of "1" from the set of values of the character χ_k within the set of values of the character χ_{k-1}. $\varphi_{k+1}^{-1}(-1)$ is a complete prototype of "-1" from the set of values of the character χ_k within the set of values of the character χ_{k+1}. It should be noted that $\chi_{\kappa-1}(g) = \left(\chi_1(g)\right)^{k-1}$ and $\chi_{\kappa+1}(g) = \left(\chi_1(g)\right)^{k+1}$. If $\chi_k(g) = 1$ then according to the equality (3.2.3) $g = 2j, j=0, 1, ..., k-1$. $P_B = 1$ exactly within the sectors with even numbers. It means that if $f(\alpha_1,...,\alpha_n) = 1$ then r_α, which satisfies the condition (3.2.5b), belongs exactly to one of these sectors. If $\chi_k(g) = -1$ then $g = 2j-1, j=0, 1, ..., k-1$. According to the equality (3.2.3) $P_B = -1$ is exactly within the sectors with odd numbers. It means that if $f(\alpha_1,...,\alpha_n) = -1$ then r_α belongs to sector that is opposite to those, to which the weighted sum is located. Exactly such sectors are described by the mapping $\varphi_{k+1}^{-1}(-1)$. It should be noted that the function P_B has the same values in opposite sectors for even values of k, and the opposite values for odd values of k (we will return below to the differences between functions P_B, which are defined with even and odd k). It follows that if Boolean function is P-realizable, then the equality (3.2.5b) is true on the entire domain of the function, and if this function is fully defined then the equality (3.2.5a) is true. Theorem is proven.

If the vector r satisfies the conditions of the Theorem 3.2.1 then we say that the vector r and its components are *acceptable* for a given function.

Remark. We will assume further on that Boolean functions fully defined. All the results for partially defined functions may be easy obtained from the results, which will be considered for the fully defined functions.

Theorem 3.2.2. The Boolean function $f(x_1,...,x_n)$ is P-realizable if and only if a acceptable vector r exists such that the following equalities are true:

$$(r \circ f, \chi_j) = 0, \ j = n+1, ..., 2^n - 1, \tag{3.2.6}$$

where χ_j, $j = n+1, ..., 2^n - 1$ are the characters of the group E_2^n, which differ from the Rademacher functions.

Proof. Necessity. Let $W = (w_0, w_1, ..., w_n)$ be a weighting vector of the function $f(x_1,...,x_n)$. According to the Theorem 3.2.1 a acceptable vector r over the field C exists such that (3.2.5a) is true: $w_0 + w_1 X_1 + ... + w_n X_n = r \circ f$. Let us decompose the function $r \circ f$ in terms of the characters of the group E_2^n:

$$r \circ f = \gamma_0 X_0 + \gamma_1 X_1 + ... + \gamma_n X_n + \gamma_{n+1} \chi_{n+1} + ... + \gamma_{2^n-1} \chi_{2^n-1}, \tag{3.2.7}$$

where $X_0, X_1, ..., X_n$ are the Rademacher functions, and $\chi_{n+1}, ..., \chi_{2^n-1}$ are all the other characters. Since such a decomposition is unique, it follows from the equalities (3.2.5) and (3.2.7) that $\gamma_0 = w_0, \gamma_1 = w_1, ..., \gamma_n = w_n$

and $\gamma_{n+1} = \gamma_{n+2} = ... = \gamma_{2^n-1} = 0$, therefore the equalities (3.2.6) are true, and vector r satisfies the conditions of the theorem.

Sufficiency. A acceptable vector r exists such that the equalities (3.2.6) are true. It follows from the equalities (3.2.7) and (3.2.6) that $\gamma_{n+1} = \gamma_{n+2} = ... = \gamma_{2^n-1} = 0$ (because the decomposition (3.2.7) is unique). But in such a case we can put $w_0 = \gamma_0$, $w_1 = \gamma_1$, ..., $w_n = \gamma_n$, and according to the Theorem 3.2.1 the function $f(x_1,...,x_n)$ is P-realizable, and $W = (w_0, w_1,..., w_n)$ is its weighting vector.

Theorem 3.2.3. If the Boolean function $f(x_1,...,x_n)$ is P-realizable with the weighting vector $W = (w_0, w_1,..., w_n)$ then the weights may be found like following:

$$w_j = (r^* \circ f, X_j), \quad j=0, 1, ..., n, \qquad (3.2.8)$$

where $r^* = (r_0^*, r_1^*, ..., r_{2^n-1}^*)$ is a acceptable solution of the system of linear algebraic equations (3.2.6) with respect to r.

Proof. If the Boolean function $f(x_1,...,x_n)$ is P-realizable then according to the Theorems 3.2.1-3.2.2 the system of linear algebraic equations (3.2.6) has a acceptable solution relatively r. Taking into account that $\gamma_{n+1} = \gamma_{n+2} = ... = \gamma_{2^n-1} = 0$ in the decomposition (3.2.7), it can be transformed to the following: $w_0 X_0 + w_1 X_1 + ... + w_n X_n = r^* \circ f$. Let us find scalar products of both parts of the last equality with the Rademacher functions:

$$\frac{1}{2^n}(r^* \circ f, X_j) = w_j', \quad j=0, 1, ..., n. \qquad (3.2.9)$$

Multiplying both parts of the equalities (3.2.9) with 2^n we obtain $w_j = 2^n w_j'$, $j = 0, 1, ..., n$. It means that the equalities (3.2.8) are true, and that the theorem is proven.

The following statement is evidently true taking into account the Theorems 3.2.1 - 3.2.3.

Theorem 3.2.4. If the Boolean function $f(x_1,...,x_n)$ is P-realizable, and the predicate P_B is defined by the equality (3.2.1) then a partially defined m-valued function $\tilde{f}(x_1,...,x_n)$, which is defined on the subset $\tilde{E}_2^n$ of the set $\tilde{E}_m^n$ and takes the values defined like follows

$$\begin{cases} \tilde{f}(\alpha_1,...,\alpha_n) = P(r_\alpha^*), \text{if } f(\alpha_1,...,\alpha_n) = 1 \\ \tilde{f}(\alpha_1,...,\alpha_n) = P(-r_\alpha^*), \text{ if } f(\alpha_1,...,\alpha_n) = -1, \end{cases}$$

is an m-valued threshold function, and any weighting vector of one of this functions is common for both functions. Here $(\alpha_1,...,\alpha_n) \in \tilde{E}_2^n \subset \tilde{E}_m^n$, $r^* = (r_0^*, r_1^*, ..., r_{2^n-1}^*)$ is a acceptable solution of the system (3.2.6), m-valued predicate P is defined by the equality (2.2.2) (k is replaced by m).

The Theorems 3.2.1-3.2.3 define a method of synthesis of the UBN, which performs a mapping described by P-realizable function f. First one should check, is the function corresponds to a P-realizable function, and if so, one should find at least one of its weighting vectors. Solving the system (3.2.6) we can check the P-realizability for a given value of m (in (3.2.1)). If it is possible to find an acceptable solution of the system (3.2.6) then a function is P-realizable, and weights may be found according to the equalities (3.2.8). Rank of the system (3.2.6) is equal to $2^n - (n+1)$. It means that $n+1 = 2^n - (2^n - (n+1))$ of unknowns are free, and we can give them arbitrary values from the acceptable ones. The other unknowns should be evaluated taking into account that they have to be acceptable. It is also possible to consider a problem from another point of view. To solve the system (3.2.6) and then to find such a value of m in (3.2.1) that the corresponding solution of the system (3.2.6) will be acceptable for a partition of the complex plane as defined by the predicate P_B.

Theorem 3.2.4 reduces the synthesis of UBN to the synthesis of MVN, which performs the mapping described by partially defined m-valued threshold function. This problem has been already considered in the Section 2.4.

We will return to the problem of synthesis in the Chapter 4, where the learning algorithms for UBN will be considered.

Let us consider some interesting features of P-realizable Boolean functions over the field of the complex numbers.

Each Boolean function $f(y_1,...,y_n)$ may be decomposed by variable y_i in Boolean alphabet $\{0,1\}$ in the following way:

$$f(y_1,...,y_n) =$$
$$= \left(y_i \& f(y_1,..., y_{i-1}, 1, y_{i+1}, y_n) \right) \vee \left(\bar{y}_i \& f(y_1,..., y_{i-1}, 0, y_{i+1}, y_n) \right)$$

We can consider decomposition by the variable y_1 for simplicity and without loss of generality. The previous property for the variable y_1:

$$f(y_1,...,y_n) = \left(y_1 \& f(1, y_2, ..., y_n) \right) \vee \left(\bar{y}_1 \& f(0, y_2, ..., y_n) \right). \quad (3.2.10)$$

The equality (3.2.10) is transformed for Boolean alphabet $\{-1, 1\}$ like follows:

$$f(x_1,...,x_n) = \left(x_1 \& f(-1, x_2,..., x_n) \right) \vee \left(-x_1 \& f(1, x_2,..., x_n) \right) \quad (3.2.11)$$

Let us denote

$$f(-1, x_2, ..., x_n) =$$
$$= f_{-1}(x_2, ..., x_n); \quad f(1, x_2, ..., x_n) = f_1(x_2, ..., x_n) \qquad (3.2.12)$$

Theorem 3.2.5. If the Boolean function $f(x_1, ..., x_n)$ is P-realizable with the weighting vector $W = (w_0, w_1, ..., w_n)$ then the function f_{-1} is P-realizable with the weighting vector $(w_0 - w_1, w_2, ..., w_n)$, and then the function f_1 is P-realizable with the weighting vector $(w_0 + w_1, w_2, ..., w_n)$.

Proof. Let $x_1 = 1$. Then the equality (3.2.11) is transformed to $f(1, x_2..., x_n) = f_1(x_2, ..., x_n)$. It means that the function f_1 is P-realizable, and its weighting vector is $(w_0 + w_1, w_2, ..., w_n)$ because $w_1 x_1 = w_1$. If $x_1 = -1$ then $w_1 x_1 = -w_1$, and taking into account the correspondence (3.2.12), the function f_{-1} is P-realizable, and its weighting vector is equal $(w_0 - w_1, w_2, ..., w_n)$.

Theorem 3.2.6. If the Boolean functions $f_a(x_2, ..., x_n)$ and $f_b(x_2, ..., x_n)$ are P-realizable with the weighting vectors $(a, w_2, ..., w_n)$ and $(b, w_2, ..., w_n)$, respectively then the function

$$f(x_1, ..., x_n) = \left(x_1 \& f_b(x_2, ..., x_n) \right) \vee \left(-x_1 \& f_a(x_2, ..., x_n) \right) \qquad (3.2.13)$$

is P-realizable with the weighting vector $W = (w_0, w_1, w_2, ..., w_n)$, where

$$w_0 = \frac{a+b}{2}; \quad w_1 = \frac{a-b}{2}. \qquad (3.2.14)$$

In the other words: if $f_a(x_2, ..., x_n)$ and $f_b(x_2, ..., x_n)$ are the P-realizable Boolean functions of n-1 variables, and their weighting vectors differ only in the 0th components then the function of n variables defined by the equality (3.2.13) also is P-realizable, and the components of its weighting vector are defined by the equalities (3.2.14).

Proof. The weighting vector, which is defined by the equalities (3.2.14) implements some function $g(x_1, ..., x_n)$. According to the equality (3.2.11) this function may be represented like

$$g = (x_1 \& g_{-1}) \vee (x_1 \& g_1). \qquad (3.2.15)$$

According to the Theorem 3.1.6 the functions of n-1 variables g_1 and g_{-1} are P-realizable with the weighting vectors $(w_0 + w_1, ..., w_n)$ and $(w_0 - w_1, ..., w_n)$ respectively. But according to (3.2.14) $w_0 - w_1 = \frac{a+b}{2} - \frac{a-b}{2} = b$; $w_0 + w_1 = \frac{a+b}{2} + \frac{a-b}{2} = a$. It means that $g_1 = f_a$ and $g_{-1} = f_b$. Comparing the equalities (3.2.13) and (3.2.15) we obtain that $f(x_1,...,x_n) = g(x_1,...,x_n)$, and conditions of the theorem are true.

Let m be fixed in the equality (3.2.1). Let Φ be a set of all the P-realizable functions of n variables. Denote Φ_1 and Φ_{-1} as the sets of the P-realizable functions of n-1 variables such that for each function $f_{-1} \in \Phi_{-1}$ a function $f_1 \in \Phi_1$ may be found (and conversely, for each $f_1 \in \Phi_1$ a function $f_{-1} \in \Phi_{-1}$ may be found) for which the weighting vectors may only differ from each other by the components w_0. If the set $\widetilde{\Phi} = \Phi_{-1} \bigcup \Phi_1$ coincides with the set of all P-realizable functions of n-1 variables then the set Φ may be obtained from the sets Φ_1 and Φ_{-1} according to the Theorems 3.2.5-3.2.6. If the set $\widetilde{\Phi}$ coincides with the set of all Boolean functions of n-1 variables then the set Φ coincides with the set of all Boolean functions of n variables. Really, suppose that such a function g exists that $g \notin \Phi$. According to the equalities (3.2.11)-(3.2.12) $g = (x_1 \& f_{-1}) \vee (-x_1 \& f_1)$. But f_1 and f_{-1} belong to $\widetilde{\Phi}$ and they are P-realizable functions. This means that according to Theorem 3.2.6 the function g also is P-realizable and $g \in \Phi$. It follows from the last considerations that if all self-dual Boolean functions or all even Boolean functions of n variables are P-realizable then all Boolean functions of n-1 variables are P-realizable.

Let us decompose the function $f(x_1,...,x_n)$ by two variables (let they will be the variables x_1, x_2 without loss of generality). Applying (3.2.11) to the functions f_1 and f_{-1} we obtain the following:

$$\begin{aligned}
f(x_1, x_2,..., x_n) &= \left(x_1 \& x_2 \& f(-1, -1, x_3,..., x_n)\right) \vee \\
&\vee \left(x_1 \& (-x_2) \& f(-1, 1, x_3,..., x_n)\right) \vee \\
&\vee \left((-x_1) \& x_2 \& f(1, -1, x_3,..., x_n)\right) \vee \\
&\vee \left((-x_1) \& (-x_2) \& f(1, 1, x_3,..., x_n)\right)
\end{aligned} \tag{3.2.16}$$

Let us denote:

$$f(-1, -1, x_3, ..., x_n) = f_{-1-1}; \quad f(-1, 1, x_3, ..., x_n) = f_{-11};$$
$$f(1, -1, x_3, ..., x_n) = f_{1-1}; \quad f(1, 1, x_3, ..., x_n) = f_{11}$$

The equality (3.2.11) may be transformed like follows: $f = \frac{1}{2}\left((1-x_1)f_{-1} + (1+x_1)f_1\right)$. On the base of the last equality we obtain from the equality (3.2.16) the following:

$$f = \frac{1}{4}((1-x_1)(1-x_2)f_{-1-1} + (1-x_1)(1+x_2)f_{-11} + \qquad (3.2.17)$$
$$+ (1+x_1)(1-x_2)f_{1-1} + (1+x_1)(1+x_2)f_{11})$$

Let $x_1 = 1$, $x_2 = 1$. Then f is reduced to f_{11}. If f is P-realizable with the weighting vector $(w_0, w_1, ..., w_n)$ then f_{11} is P-realizable function of n-2 variables and evidently, its weighting vector is $(w_0 + w_1 + w_2, w_3, ..., w_n)$. It is possible to prove in a similar way that the functions $f_{1-1}, f_{-11}, f_{-1-1}$ are also P-realizable. The following lemma has been proven by the last considerations.

Lemma 3.2.1. If the Boolean function $f(x_1, ..., x_n)$ is P-realizable with the weighting vector $(w_0, w_1, ..., w_n)$ then the functions $f_{11}, f_{1-1}, f_{-11}, f_{-1-1}$ of n-1 variables are also P-realizable with the weighting vectors $(a_{ij}, w_3, ..., w_n)$, $i, j \in \{1, -1\}$, and

$$a_{11} = w_0 + w_1 + w_2; a_{1-1} = w_0 + w_1 - w_2; \qquad (3.2.18)$$
$$a_{-11} = w_0 - w_1 + w_2; a_{-1-1} = w_0 - w_1 - w_2.$$

Theorem 3.2.7. If some four Boolean functions of n-2 variables $x_3, ..., x_n$ $f_{11}, f_{1-1}, f_{-11}, f_{-1-1}$ are P-realizable with the weighting vectors $(a_{ij}, w_3, ..., w_n)$, $i, j \in \{1, -1\}$ respectively (which differ from each other may be only by the components a_{ij}) then the function $f(x_1, ..., x_n)$ defined by the equality (3.2.16) is also P-realizable with the weighting vector $(w_0, w_1, w_2, w_3 ..., w_n)$ where

$$w_0 = \frac{1}{4}(a_{11} + a_{1-1} + a_{-11} + a_{-1-1});$$

$$w_1 = \frac{1}{4}(a_{11} + a_{1-1} - a_{-11} - a_{-1-1}); \qquad (3.2.19)$$

$$w_2 = \frac{1}{4}(a_{11} - a_{1-1} + a_{-11} - a_{-1-1}).$$

The proof of this theorem is analogous to proof of the Theorem 3.2.6, and we leave it to the reader.

Applying the equality (3.2.11) sequentially to all the variables, it is possible to decompose the function $f(x_1, ..., x_n)$ by all the variables. The functions $f_{1...1}, f_{1...1-1}, ..., f_{-1...-1}$ will be the values of the function f on all 2^n combinations of variables. The corresponding analogy of the equality (3.2.16) will become a disjunctive canonical form of the function f. For example, for n=2

$$f_{11} = f(1, 1), f_{1-1} = f(1, -1), f_{-11} = f(-1, 1), f_{-1-1} = f(-1, -1).$$

The analogy of the equality (3.2.16) will be the following:

$$f(x_1, x_2) = (x_1 \,\&\, x_2 \,\&\, f(-1, -1)) \vee (x_1 \,\&\, (-x_2) \,\&\, f(-1, 1)) \vee$$
$$\vee ((-x_1) \,\&\, x_2 \,\&\, f(1, -1)) \vee ((-x_1) \,\&\, (-x_2) \,\&\, f(1, 1)).$$

If the function $f(x_1, x_2)$ is P-realizable then according to the Lemma 3.2.1 the functions $f_{11}, f_{1-1}, f_{-11}, f_{-1-1}$ will be also P-realizable. Since these functions are constants, their weighting vectors contain only the components w_0: $a_{11}, a_{1-1}, a_{-11}, a_{-1-1}$. Let us add together the equalities (3.2.19):

$$w_0 + w_1 + w_2 = \frac{1}{4}(3a_{11} + a_{1-1} + a_{-11} - a_{-1-1}).$$ On the other hand according

to the (3.2.18) $\qquad a_{11} = w_0 + w_1 + w_2,$ $\qquad$ therefore

$$a_{11} = \frac{1}{4}(3a_{11} + a_{1-1} + a_{-11} - a_{-1-1}) \qquad \text{from} \qquad \text{which}$$

$a_{11} - a_{1-1} - a_{-11} + a_{-1-1} = 0.$ The last equality may be written also in the following form:

$$(a, \chi_3) = 0, \tag{3.2.20}$$

where $a = (a_{11}, a_{1-1}, a_{-11}, a_{-1-1})$, $\chi_3 = (1, -1, -1, 1)$ is a single character of the group E_2^n which does not coincide with any Rademacher function. Let us compare the equality (3.2.20) with the equality (3.1.12) to which the system (3.2.6) is reduced for $n=2$. Such a comparison immediately shows that

$$a = r \circ f, \tag{3.2.21}$$

where vector r is a solution of the equation (3.1.12). Similar considerations for the functions of n variables lead to the following result:

$$(a, \chi_j) = 0, \quad j=n+1, ..., 2^n - 1, \tag{3.2.22}$$

where $a = (a_{11...1}, ..., a_{-1-1...-1})$. Comparing the systems (3.2.6) and (3.2.22) we obtain the equality (3.2.21). It means that the components of the vector a are equal to the values of the weighted sum on the corresponding values of variables, and the systems of the equations (3.2.6) and (3.2.22) are equivalent.

If k in the (3.2.1) is an even number, then the corresponding predicate P_B is called *k-even* predicate. If k in the (3.2.1) is an odd number then the corresponding predicate P_B is called *k-odd* predicate [Aizenberg I. (1991)].

Proving Theorem 3.2.1 we have seen that *k-even* and *k-odd* predicates P_B have one principle difference. For a *k-even* predicate its values in mutual opposite sectors, on which it separates the complex plane, are always the same. For a k-odd predicate its values in mutual opposite sectors are always opposite. This fact is illustrated in Fig. 3.2.2.

The interesting feature of the k-odd predicate is that it ensures that all Boolean threshold functions are P-realizable with their real-valued weighting vectors. Indeed, for any k-odd predicate

$$P_B(z) = \begin{cases} 1, \text{ if } \arg z = 0 \\ -1, \text{ if } \arg z = \pi, \end{cases}$$

but the argument of the real-valued weighted sum for any Boolean threshold function is always equal 0 or π (see also Fig. 3.2.2b with the example for $k=3$, $m=6$). It is interesting that some important features of threshold Boolean functions considered for example in [Chow (1961ab)] [Dertouzos (1965)], [Muroga (1971)] may be generalized on the P-realizable Boolean functions (in the case of k-odd predicate the group classification considered in the mentioned papers and books may be completely generalized).

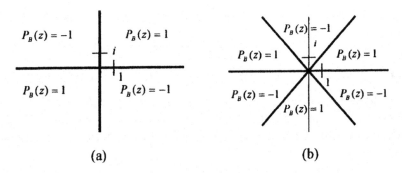

(a)   (b)

Figure 3.2.2. (a) - the example of k-even predicate ($k=2$) in (3.2.6);
(b) - the example of k-odd predicate ($k=3$ in (3.2.6))

Lemma 3.2.1. If the Boolean function $f(x_1, ..., x_i, ..., x_n)$ is P-realizable with the weighting vector $W = (w_0, w_1, ..., w_i, ..., w_n)$ then the function $f' = f(x_1, ..., -x_i, ..., x_n)$ is also P-realizable with weighting vector $W' = (w_0, w_1, ..., -w_i, ..., w_n)$.

Proof. According to the equalities (3.2.8) $w'_j = (r \circ f, X_j) = w_j, \ j \neq i$, where r is a acceptable solution of the system (3.2.6) for function f. $w'_i = (r \circ f, -X_j) = -(r \circ f, X_j) = -w_i$,   and   it   means   that $W' = (w_0, w_1, ..., -w_i, ..., w_n)$ is a weighting vector of the function f'.

Lemma 3.2.2. If the Boolean function $f(x_1, ..., x_i, ..., x_j, ..., x_n)$ is P-realizable with the weighting vector $W = (w_0, w_1, ..., w_i, ..., w_j, ..., w_n)$

then the function $f' = f(x_1, ..., x_j, ..., x_i, ..., x_n)$ is P-realizable with the weighting vector $W' = (w_0, w_1, ..., w_j, ..., w_i, ..., w_n)$.

Proof. Evidently, $w'_k = w_k$ for $k \neq j$. $w'_i = (r \circ f, X_j) = w_j$ and $w'_j = (r \circ f, X_i) = w_i$, where r is a acceptable solution of the system (3.2.6) for function f. It means that $W' = (w_0, w_1, ..., w_j, w_i, ..., w_n)$ is a weighting vector of the function f'.

Lemma 3.2.3. If the Boolean function $f(x_1, ..., x_{i-1}, x_i, x_{i+1}, ..., x_n)$ is P-realizable with the weighting vector $W = (w_0, w_1, ..., w_n)$ then the function $f' = f(-x_1, ..., -x_n)$ (an opposite function to f) is P-realizable with the weighting vector $W' = (w_0, -w_1, ..., -w_n)$.

The Lemma 3.2.3 follows directly from the Lemma 3.2.1.

The following three lemmas are only valid for the functions, which are P-realizable for k-odd predicates.

Lemma 3.2.4. If the Boolean function $f(x_1, ..., x_{i-1}, x_i, x_{i+1}, ..., x_n)$ is P-realizable with the weighting vector $W = (w_0, w_1, ..., w_{i-1}, w_i, w_{i+1}, ..., w_n)$ then the function $f' = x_i f(x_1 x_i, ..., x_{i-1} x_i, x_i, x_{i+1} x_i, ..., x_n x_i)$ (obtained by equidualisation [Dertouzos (1965)] from the function f) is P-realizable with the weighting vector $W' = (w_i, w_1, ..., w_{i-1}, w_0, w_{i+1}, ..., w_n)$.

Proof. Let obtain the weighting vector for the function f'. According to the equalities (3.2.8) we obtain the following:

$$w'_0 = (r \circ f \circ X_i, X_0) = (r \circ f, X_i \circ X_0) = (r \circ f, X_i) = w_i;$$
$$w'_j = (r \circ f \circ X_i, X_i \circ X_j) = (r \circ f, X_i \circ X_i \circ X_j) =$$
$$= (r \circ f, X_j) = w_j, \ j \neq 0, i;$$
$$w'_i = (r \circ f \circ X_i, X_i) = (r \circ f, X_i \circ X_i) = (r \circ f, X_0) = w_0.$$

It means that $W' = (w_i, w_1, ..., w_{i-1}, w_0, w_{i+1}, ..., w_n)$ is a weighting vector of the function f'.

Lemma 3.2.5. If the Boolean function $f(x_1, ..., x_n)$ is P-realizable with the weighting vector $W = (w_0, w_1, ..., w_n)$ then the function $f' = -f(x_1, ..., x_n)$ (negation of the function f) is P-realizable with the weighting vector $W' = (-w_0, -w_1, ..., -w_n)$.

Proof. If the vector r is a acceptable solution of the system (3.2.6) for the function f then it will be also a acceptable solution of the same system for the function $-f$ (this fact is followed from the Theorem 3.2.1 taking into account also that the predicate P_B is k-odd). According to the equalities (3.2.8) we obtain   the   following:

$$w_j' = (r \circ (-f), X_j) = -(r \circ f, X_j) = -w_j, \quad j = 0, 1, ..., n.$$

Lemma 3.2.6. If the Boolean function $f(x_1, ..., x_n)$ is P-realizable with the weighting vector $W = (w_0, w_1, ..., w_n)$ then the function $f' = -f(-x_1, ..., -x_n) = f^D$ ($f' = f^D$ is a function dual to the function f) is P-realizable with the weighting vector $W' = (-w_0, w_1, ..., w_n)$.

Proof. If the vector r is a acceptable solution of the system (3.2.6) for the function f, it will be also a acceptable solution of the same system for the function f^D (this fact follows from the Theorem 3.2.1 taking into account also that predicate P_B is k-odd). According to the equalities (3.2.8) we obtain   the   following: $w_0' = (r \circ (-f), X_0) = -(r \circ f, X_0) = -w_0;$

$$w_j' = \left(r \circ (-f), (-X_j) \right) = (r \circ f, X_j) = w_j, \quad j = 1, ..., n.$$

Remark. The Lemmas 3.2.4-3.2.6 are not true for k-even predicates P_B because the vector r in such a case is not a acceptable solution of the system (3.2.6) simultaneously for the functions f and f'. It is easy to derive some insight from the Theorem 3.2.1. If the function f is P-realizable with k-odd predicate then the vector r, which is a acceptable solution of the system (3.2.6) for the function f, will be a acceptable solution of the same system for its negation the function $-f$ and for any function obtained from f by changing of the sign of any of its values. The situation is absolutely different for k-even predicate (see Theorem 3.2.1). If $f'(\alpha_1, ..., \alpha_n) = -f(\alpha_1, ..., \alpha_n)$ then, evidently, $r_\alpha' = ir_\alpha$ or $r_\alpha' = -ir_\alpha$, where i is imaginary unity. The last consideration shows that it is impossible to prove an analog of the Lemma 3.2.4 for k-even predicates, but the following two lemmas are the analogs of the Lemmas 3.2.5 and 3.2.6 for k-even predicate P_B.

Lemma 3.2.7. If the Boolean function $f(x_1, ..., x_n)$ is P-realizable with the weighting vector $W = (w_0, w_1, ..., w_n)$ and k-even predicate P_B then the function $f' = -f(x_1, ..., x_n)$ (negation of the function f) is P-realizable with the weighting vector $W' = (iw_0, iw_1, ..., iw_n)$, where i is imaginary unity.

Proof. If the vector r is a acceptable solution of the system (3.2.6) for the function f then the vector $-ir$ is a acceptable solution of the same system for the function $-f$ (this fact follows from the Theorem 3.2.1 taking into account also that the predicate P_B is k-even). According to the equalities (3.2.8) we obtain the following:

$$w'_j = (-ir \circ (-f), X_j) = i(r \circ f, X_j) = iw_j, \quad j = 0, 1, ..., n.$$

Lemma 3.2.8. If the Boolean function $f(x_1, ..., x_n)$ is P-realizable and its weighting vector is $W = (w_0, w_1, ..., w_n)$, then the function $f' = -f(-x_1, ..., -x_n) = f^D$ ($f' = f^D$ is a function dual to the function f) is P-realizable with the weighting vector $W' = (iw_0, -iw_1, ..., -iw_n)$.

Proof. If the vector r is a acceptable solution of the system (3.2.6) for the function f then the vector $-ir$ is a acceptable solution of the same system for the function f^D (this fact follows from the Theorem 3.2.1 taking into account also that the predicate P_B is k-even). According to the equalities (3.2.8) we obtain the following:

$$w'_0 = (-ir \circ (-f), X_0) = i(r \circ f, X_0) = iw_0;$$

$$w'_j = \left(-ir \circ (-f), (-X_j)\right) = -i(r \circ f, X_j) = -iw_j, \quad j = 1, ..., n.$$

Definition 3.2.3. A *characteristic vector* of Boolean function $f(x_1, ..., x_n)$ of n variables is a vector $B = (b_0, b_1, ..., b_n)$ which is defined like follows: $b_j = (f, X_j)$, $j = 0, 1, ..., n$, i.e. components of the vector B are the scalar products of the vector of function values with the corresponding Rademacher functions.

One may compare the Definition 3.2.3 with the analogous definitions given in [Chow (1961ab)] [Dertouzos (1965)], [Muroga (1971)].

If the characteristic vectors of some functions contain the same components (without taking into account their order and signs) such vectors are called *equivalent* and the corresponding functions belong to the same equivalence class. The characteristic vectors of Boolean threshold functions are always different [Dertouzos (1965)]. The characteristic vectors of all functions from the same equivalence class may be easily derived from each other. The following lemma shows, how this can be done.

Lemma 3.2.9. If the Boolean function f' is obtained from the Boolean function f in one of the following ways: 1) changing the sign (negation) of the variable x_i; 2) permutation of the variables x_i, x_j; 3) equidualisation of the variable x_i; 4) $f' = -f$; 5) $f'(x_1, ..., x_n) = -f(-x_1, ..., -x_n) = f^D$ ($f' = f^D$ is a dual function to f) then the characteristic vectors $B = (b_0, b_1, ..., b_n)$ and $B' = (b'_0, b'_1, ..., b'_n)$ of the functions f and f' respectively are equivalent and the vector B' is equal (respectively, for the mentioned five cases): 1) $(b_0, ..., b_{i-1}, -b_i, b_{i+1}, ..., b_n)$; 2) $(b_0, ..., b_j, ..., b_i, ..., b_n)$; 3) $(b_i, ..., b_{i-1}, b_0, b_{i+1}, ..., b_n)$; 4) $(-b_0, -b_1, ..., -b_n)$; 5) $(-b_0, b_1, ..., b_n)$.

Proof. According to the Definition 3.2.3 of a characteristic vector we obtain the following for the components of the vector $B' = (b'_0, b'_1, ..., b'_n)$ (each of five cases is considered separately):

1) $b'_i = (f, -X_i) = -(f, X_i) = -b_i$;

2) $b'_i = (f, X_j) = b_j$; $b'_j = (f, X_i) = b_i$;

3) $\begin{cases} b'_0 = (f \circ X_i, X_0) = (f, X_i \circ X_0) = (f, X_i) = b_i; \\ b_i = (f \circ X_i, X_i) = (f, X_i \circ X_i) = (f, X_0) = b_0; \\ b_j = (f \circ X_i, X_i \circ X_j) = (f, X_i \circ X_i \circ X_j) = (f, X_j) = b_j; j \neq 0, i \end{cases}$

4) $b'_j = (-f, X_j) = -(f, X_j) = b_j$, $j = 0, 1, ..., n$;

5) $\begin{cases} b'_0 = (-f, X_0) = -(f, X_0) = b_0; \\ b'_j = (-f, -X_j) = (f, X_j) = b_j \quad j = 0, 1, ..., n. \end{cases}$

Lemma 3.2.9 defines five invariant operations over the characteristic vectors. Such operations make it possible to obtain an equivalent characteristic vector from another one.

The following important theorem follows directly from the Lemmas 3.2.1 - 3.2.6 and 3.2.9 for P-realizable Boolean functions with k-odd predicates (one may compare with the similar results proven in [Dertouzos (1965)] only for the threshold functions. It should be also noted that proofs without using of the characters features even for the threshold functions are much more complicated).

Theorem 3.2.8. The set of the P-realizable Boolean functions of n variables with the given k-odd predicate P_B is separated onto equivalence classes of according to their characteristic vectors. The characteristic and weighting vectors of the functions belonging to the same class are obtained each from other by any permutation of the components, or by changing the sign (negation) of the component (components). A permutation or changing the sign of components of the characteristic vector respects a permutation or changing the sign of corresponding components of the weighting vector.

The next theorem follows directly from the Lemmas 3.2.1-3.2.3, 3.2.7-3.2.8 and 3.2.9 for P-realizable Boolean functions with k-even predicates.

Theorem 3.2.9. If the Boolean function $f(x_1,...,x_n)$ with the characteristic vector $B = (b_0, b_1, ..., b_n)$ is P-realizable with the weighting vector $W = (w_0, w_1, ..., w_n)$ and k-even predicate P_B, then the characteristic vectors obtained from the vector B by any permutations of the components $b_1, ..., b_n$, or changing the sign (negation) of the any component also correspond to P-realizable Boolean functions. The weighting vectors for these functions may be obtained according to the

Lemmas 3.2.1-3.2.3 and 3.2.7-3.2.8 depending of transformation of the characteristic vector.

The following theorem establishes a connection between the characteristic vectors of the functions of n and n-1 variables.

Theorem 3.2.10. If the Boolean function $f(x_1,...,x_n)$ is connected with Boolean functions $f_a(x_2,...,x_n)$ and $f_b(x_2,...,x_n)$ by the equality (3.2.13)

$$f(x_1,...,x_n) = (x_1 \& f_b(x_2, ..., x_n)) \vee (-x_1 \& f_a(x_2, ..., x_n))$$ then the

characteristic vectors B, B^a, B^b of the functions f, f_a, f_b are connected by the following relation:

$$b_0 = b_0^a + b_0^b; \quad b_1 = b_0^a - b_0^b; \quad b_j = b_{j-1}^a + b_{j-1}^b, j = 2, ..., n. \qquad (3.2.23)$$

Proof. It follows from the equalities (3.2.13) and (3.2.11) that $f_a(x_2,...,x_n) = f(1,x_2,...,x_n)$ and $f_b(x_2,...,x_n) = f(-1,x_2,...,x_n)$. The following relations between the vectors of values of the functions f, f_a, f_b follow from the last two equalities:

$$f^{(j)} = \begin{cases} f_a^{(j)}, & j = 0, 1, ..., 2^{n-1} - 1; \\ f_b^{(j-2^{n-1})}, & j = 2^{n-1}, ..., 2^n - 1 \end{cases}$$

It should be noted that the characters of the groups E_2^n and E_2^{n-1}, which are the corresponding Rademacher functions are connected by the following equalities:

$$X_j^n = (X_{j-1}^{n-1} \mid X_{j-1}^{n-1}), \ j = 2, 3, ..., n;$$

$$X_0^n = (X_0^{n-1} \mid X_0^{n-1}), \ X_1^n = (\underbrace{1,1,...,1}_{2^{n-1} \text{ times}} \underbrace{-1,-1,...,-1}_{2^{n-1} \text{ times}}),$$

where $\mid$ is a symbol of concatenation of the 2^{n-1}- dimensional vectors into the 2^n-dimensional one. Taking into account the last equalities and the Definition 3.2.3 of the characteristic vector it is easy to verify that the equalities (3.2.23) are true.

To illustrate a classification of *P*-realizable Boolean functions defined by their characteristic vectors and Theorems 3.2.8 - 3.2.9, let us consider examples for the functions of 2 and 3 variables. The functions of two variables are split into three equivalence classes (here and further we will present a class by one of the equivalent characteristic vectors, exactly the vector with non-increased and non-negative components): 1) (4, 0, 0); 2) (2, 2, 2); 3) (0, 0, 0). The classes 1 (6 functions) and 2 (8 functions) are established only by threshold functions, the class 3 - by XOR and not-XOR functions which are not threshold. The functions of three variables are split into seven equivalence classes: 1) (8, 0, 0, 0); 2) (6, 2, 2, 2); 3) (4, 4, 4, 0); 4) (0, 0, 0, 0); 5) (2, 2, 2, 2); 6) (4, 4, 0, 0); 7) (4, 0, 0, 0). The classes 1-3 contain all 104 threshold functions, the classes 4-7 - all the non-threshold functions.

According to the Theorems 3.2.8 - 3.2.9 the problem of the seeking for weighting vectors for all the P-realizable functions of a given number of variables may be reduced to the solution of this problem only for representatives of the classes. The weighting vectors for all the other functions may be obtained by changing the sign of the weights and their permutation. An asymptotic estimate of number of the equivalence classes may be obtained from the Definition 3.2.3 of the characteristic vector (classes are defined exactly by these vectors) and is equal for $n>3$ to

$$q \to \frac{2^{2^n}}{(n+1)! 2^{n+1}} \text{ [Aizenberg I. (1991)]}.$$

3.   A GLOBAL PROBLEM OF WEIGHTING DESCRIPTION OF BOOLEAN FUNCTIONS. *P*-REALIZABLE BOOLEAN FUNCTIONS OVER THE RESIDUE CLASS RING

We have considered in the previous two sections the P-realizable Boolean functions over the field of the complex numbers. It was shown that the number of the Boolean functions of n variables represented with $n+1$ weights is considerably large in comparison with the number of threshold Boolean functions. This result is achieved using a new basic set for weights and a new type of the activation function. Let us expand the main features of P-realizable Boolean functions considered above on the more general case (we mean different basic sets for weights, different activation functions, and different number of weights). We will use the results presented in [Aizenberg N. et al. (1988), (1989)], [Aizenberg I. (1985)], [Aizenberg & Aizenberg (1991)] with the more recent supplements.

Let A be some algebra (e.g., residue class ring, finite field, the field of the complex numbers, etc.). Assume that among the elements of this algebra there are two elements 0 and 1, the properties of which at this stage are not specialized. $W = (w_0, w_1, ..., w_l)$ is a vector with the components in A, $Y = (y_1, ..., y_n)$ is a vector whose components are the elements 0 and 1, $F = F(W, Y)$ is a function defined on the elements of the algebra A and taking values also in A (the weighting function), P is a two-valued predicate defined on A.

Definition 3.3.1. The Boolean function $f(y_1, ..., y_n)$ is *PW*-realizable (or *PW*-described [Aizenberg N. et al. (1988)]) over the given algebra A, if the equality

$$P[F(W,Y)] = f(y_1, ..., y_n) \qquad (3.3.1)$$

is true for the whole domain of the function f.

We say that Boolean functions of a given number of variables have a complete description over the algebra A, if for each function there exists a describing weighting vector for some F and P. Investigation of the general techniques of the description of Boolean functions like (3.3.1) is of immediate practical importance, because this description provides a mathematical model of the operation of the neural element (NE) realizing the given function. In this model, vectors over the set $\{0, 1\}$ describe the set of input signals, the algebra A describes the internal operating environment of the NE, the function F describes the NE operating mechanism, the weighting vector W converts the input signals to values from A, and the predicate P converts the result produced by the NE into an output signal.

The specific form of the logic function $f(y_1, ..., y_n)$ realized by the given NE is determined by the combination of weights W, the form of the function F and the predicate P. If none of W, F, P can be altered, the given NE realizes precisely one Boolean function. If every Boolean function of a given number of variables can be realized by variation of W, F, P the NE is called *universal* (universal binary neuron - UBN).

As it was shown above that there are no difficulties in the construction of UBN over the field of the complex numbers. In general, a practical solution of this problem may be sensitive to the particular choice of the basic algebra, the function F, and the predicate P.

For UBN, we have the following interrelationship between the number ω of possible weight combinations from W, the number φ of functions $F(W, Y)$ describing different operating modes, and the number ρ of predicates P, which may be set on the output of the UBN:

$$\omega\varphi\rho \geq 2^{2^n}. \qquad (3.3.2)$$

If there is a finite number η of elements in the algebra A, and each of $l+1$ weights may take any value from A independently of the other weights, this formula takes the form

$$\eta^{l+1}\varphi\rho \geq 2^{2^n}. \qquad (3.3.3)$$

The relationships (3.3.2) and (3.3.3) are necessary, but insufficient conditions of universality for the multifunctional NE. Thus, for instance for the threshold element, the weighting function over the field of the real numbers and the condition (3.3.2) holds, but a threshold neuron is not universal. On the other hand, we can describe a mathematical model of a

UBN for which (3.3.2) is satisfied as equality. The synthesis problem of UBN reduces to finding, for a given Boolean function $f(y_1, ..., y_n)$ and a given algebra A, the weighting vector W, the function $F(W, Y)$, and the predicate P subject to various constraints.

Let A be an algebra with respect to some operations. Assume that these operations include the binary operation "+", relative to which the set of elements of the algebra A forms an additive Abelian group, and the element designated by the symbol 0 is the zero element of the group. Also assume that they include the operation " • ", relative to which the set of nonzero elements of the algebra A forms a multiplicative group, in which the element designated by the symbol 1 is the identity. In what follows, we only consider algebras with these properties. As the first restriction on the function F, we assume that it is linear in W, i.e., we only consider functions $F(W, Y)$ having the form

$$F(W,Y) = w_0 F_0(Y) + w_1 F_1(Y) + ... + w_l F_l(Y). \qquad (3.3.4)$$

In what follows we assume that all functions $F_0, F_1, ..., F_l$ are polynomials in $y_1, ..., y_n$. If all $y_1, ..., y_n$ take only the values 0 and 1, we may take each polynomial $F_i(Y)$ to have the form

$$F_i(Y) = a_0^{(i)} + a_1^{(i)} y_n + a_2^{(i)} y_{n-1} + a_3^{(i)} y_{n-1} x_n + a_{2^n-1}^{(i)} y_1 y_2 ... y_n;$$
$$a_j^{(i)} \in A, j = 0, 1, ..., 2^n - 1; \ i = 0, 1, ..., l. \qquad (3.3.5)$$

Substituting (3.3.5) in (3.3.4), expanding the parentheses and collecting similar terms, we obtain

$$F(W,Y) = v_0 + v_1 y_n + v_2 y_{n-1} + v_3 y_{n-1} y_n +$$
$$+ v_{\alpha_0 + 2\alpha_1 + ... + 2^{n-1}\alpha_{n-1}} y_1^{\alpha^n-1} ... y_n^{\alpha_0} + ... + v_{2^n-1} y_1 y_2 ... y_n; \qquad (3.3.6)$$
$$\alpha_i \in \{0,1\}; i = 0, 1, ..., n-1.$$

In the last expression, all v_s, $s = 0, 1, ..., 2^n - 1$ are linear combinations of the weights $w_0, w_1, ..., w_l$ and the coefficients $a_j^{(i)}$, $i = 0, 1, ..., l$; $j = 0, 1, ..., 2^n - 1$. All v_s clearly belong to the algebra A.

Let us introduce the following functions of the variables $y_1, ..., y_n$: the function identically equal to the constant 1 (zero-rank conjunction); functions equal to the variables $y_1, ..., y_n$ (first-rank conjunctions); functions equal to all possible products of two variables (second-rank conjunctions), and so on up to the function equal to the product of all $y_1, ..., y_n$ (n^{th} rank conjunction). Consider the case when all $F_i(Y)$ are selected from this set of functions. In this case, we have a polynomial weighting function of the form (3.3.6): $v_{i_0} = w_0$, $v_{i_1} = w_1$, ..., $v_{i_l} = w_l$, in which $l+1$ coefficients may take nonzero values and the remaining $2^n - l - 1$ coefficients may take only zero values. This weighting description of Boolean functions will be called *polynomial*.

Consider the conjunctive (Reed-Muller) transformation matrix K over the algebra A, proposed in [Aizenberg N. (1978)], [Aizenberg N. & Trofimluk (1981)]:

$$K_n = \begin{pmatrix} 1 & 0 \\ 1 & 1 \end{pmatrix}^{\otimes n},\qquad (3.3.7)$$

where $\otimes n$ is the symbol of the Kronecker n^{th} degree. The elements in the columns of this matrix are the values of the elementary conjunctions of all ranks, ordered lexicographically. We can use the matrix K to obtain the expansion of the Boolean function of n variables in elementary conjunctions, i.e., the coefficients of its modulo 2 polynomial. The matrix K is inverse to itself over the field $GF(2)$, and has the inverse

$$K = \begin{pmatrix} 1 & 0 \\ -1 & 1 \end{pmatrix}^{\otimes n}$$

over the field of the real numbers.

Denoting by $\widetilde{V}$ the column vector with the coefficients of the polynomial weighting function, $\widetilde{V} = (v_0, v_1, ..., v_{2^n-1})^T$, and by $\widetilde{F}$ the column vector with the values of the function $F(V, Y)$ on various combinations of variable values,

$$\widetilde{F} = \left(F(V,(0,0,...,0)), F(V,(0,0,...,0,1)), F(V,(0,0,...,1,0)), ..., F(V,(1,...,1)) \right)^T,$$

we may write $K\widetilde{V} = \widetilde{F}$. Denoting by f the column of values of the function $f(y_1, ..., y_n)$, we obtain

$$P\left[K\widetilde{V} \right] = f,\qquad (3.3.8)$$

where the action of the predicate P on a column vector is defined in the sense of the action of the predicate P on each element of the vector column. The equation (3.3.8) is basic for polynomial weighting descriptions of Boolean functions. It enables us to determine, which particular function $f(y_1,...,y_n)$ specified by the column of values f, is realized by the given function F on the given predicate P. From (3.3.8) we can determine, what values the predicate P should take for the elements from $\widetilde{F}$, if we have to realize the function $f(y_1,...,y_n)$ by a given polynomial $F(V,Y)$: $P[\widetilde{F}] = f$. Let us take some function $f(y_1,...,y_n)$ whose values form the vector-column f. Construct the vector-column

$p = (p_0, p_1, ..., p_{2^n-1})^T$ from the elements of A corresponding to the condition $P(p_i) = f_i$, $i = 0, 1, ..., 2^n - 1$. Treating the elements of the vector p as the values of some weighting polynomial realizing the given Boolean function, we may write $K\tilde{V} = p$, whence, using the symbolic notation $p = P^{-1}(f)$, we obtain $\tilde{V} = K^{-1}p = K^{-1}P^{-1}(f)$. Forming the vector p in all possible ways from the elements of the algebra A satisfying this condition, we obtain all the functions $F(V, Y)$ describing the given Boolean function for the given predicate P.

The simplest case of polynomial weighting (parametric) description of Boolean function corresponds to the situation, when the weighting function is linear in both W and Y:

$$F(W, Y) = w_0 + w_1 y_1 + ... + w_n y_n . \qquad (3.3.9)$$

Evidently, the substitution of the (3.3.9) into (3.3.1) leads to the Definition 3.1.2 of P-realizable Boolean function and the equality (3.1.5). Exactly this case is the most interesting from the point of view of neural technique which usually deals with the representation of function by $n + 1$ weights.

Let us consider some important features of P-realizable Boolean functions over the algebra A, which are common for any algebra (one may compare with features of the of P-realizable Boolean functions over the field of the complex numbers considered in the Section 3.2). The following several lemmas are true (we will not consider the proofs of Lemmas 3.3.1 - 3.3.5 here, but the interested reader can prove these as an exercise).

Lemma 3.3.1. (Interchanging of arguments). If the Boolean function $f(y_1, ..., y_i, y_j, ..., y_n)$ is P-realizable over the algebra A then the function $f' = (y_1, ..., y_j, y_i, ..., y_n)$ is also P-realizable over the algebra A.

Lemma 3.3.2. (Negation of argument). If the Boolean function $f(y_1, ..., y_i, ..., y_n)$ is P-realizable over the algebra A, which have a structure of additive Abelian group, then the function $f' = (y_1, ..., \bar{y}_i, ..., y_n)$ is also P-realizable over the algebra A.

A Boolean function of n variables $f(y_1, ..., y_n)$ is said to be of the same type (belongs to the same class) as the Boolean function of n variables $g(y_1, ..., y_n)$ (PN-classification, [Dertouzos (1965)]) if $f(y_1, ..., y_n)$ is transformed to $g(y_1, ..., y_n)$ by some interchanging of arguments, or replacement of some arguments by their negations.

Lemma 3.3.3. (Connection between P-realization of the functions of different number of variables). If the Boolean function of n variables $f(y_1, ..., y_{i-1}, y_i, y_{i+1}, ..., y_n)$ is P-realizable over the algebra A, then the functions of n-1 variables

$$f_{i0}(y_1, ..., y_{i-1}, y_{i+1}, ..., y_n) = f(y_1, ..., y_{i-1}, 0, y_{i+1}, ..., y_n),$$
$$f_{i1}(y_1, ..., y_{i-1}, y_{i+1}, ..., y_n) = f(y_1, ..., y_{i-1}, 1, y_{i+1}, ..., y_n),$$

for which $f(y_1, ..., y_{i-1}, y_i, y_{i+1}, ..., y_n) = \bar{y}_i f_{i0} \vee y_i f_{i1}$, are also P-realizable over the algebra A and among the weighting vectors of the functions f_{i0} and f_{i1} there are vectors differing only in the weight w_0.

Conversely, if for some predicate P over an arbitrary algebra A, two Boolean functions of n-1 variables f_{i0} and f_{i1} are P-realizable and have weighting vectors that differ only in the weight w_0, then the Boolean function of n variables, which is the composition of the functions f_{i0}, f_{i1} also is P-realizable.

If for some given $t \in A$, for any element $z \in A$ we have the equality $P(t - z) = 1 - P(z)$ we will call such a predicate P the t-odd (or simple *odd*) (one may compare with k-odd predicate over the field of the complex numbers considered in the Section 3.2). The following two lemmas are true only for functions, which are P-realizable with odd predicate over some algebra A.

Lemma 3.3.4. If the Boolean function $f(y_1, ..., y_n)$ is P-realizable over the algebra A then the function $f' = \bar{f}(\bar{y}_1, ..., \bar{y}_n) = f^D$ ($f' = f^D$ is a function dual to the function f) is also P-realizable over the algebra A.

Lemma 3.3.5. If the Boolean function $f(y_1, ..., y_{i-1}, y_i, y_{i+1}, ..., y_n)$ is P- realizable over the algebra A then the function $f' = y_i \oplus f(y_1 \oplus y_i, ..., y_{i-1} \oplus y_i, y_i, y_{i+1} \oplus y_i, ..., y_n \oplus y_i)$ (obtained by equidualisation [Dertouzos (1965)] from the function f) is also P-realizable over the algebra A.

Boolean functions of n variables $f(y_1, ..., y_n)$ and $g(y_1, ..., y_n)$ are in the same equivalence class if $f(y_1, ..., y_n)$ may be transformed to $g(y_1, ..., y_n)$ by applying the transformations of interchanging of arguments, negation of arguments, passage to dual function, equidualisation $g(y_1, ..., y_n)$ (SD-classification, [Dertouzos (1965)]). Evidently that Lemmas 3.3.1-3.3.5 cover the classification of Boolean functions by characteristic vectors described in the Section 3.2.

Since the P-realizable Boolean functions over the field of the complex numbers have been considered in the Sections 3.1-3.2, we will concentrate now on the P-realizable Boolean functions over the finite algebras. Let us start from the P-realization of Boolean functions over the residue class ring.

Let $A = Z_m$, where Z_m is the residue class ring mod m. We will investigate two problems here: 1) determine the minimum cardinality of the ring, over which all Boolean functions of n variables are P-realizable by a single predicate and find this predicate that will be an activation function of corresponding UBN; 2) determine the minimum cardinality of the ring over which all Boolean functions of n variables are P-realizable by predicates, which are possibly different for different functions. The first problem is solvable for any number of variables n.

Theorem 3.3.1. For any number of variables n, a value of m exists such that all Boolean functions of n variables are P-realizable over the ring Z_m with the same predicate.

Proof. Consider the ring Z_m for $m \geq 2^n 2^{2^n}$. Take an arbitrary Boolean function $f(Y)$. Choose $w_n = 1$, $w_{n-1} = 2$, ..., $w_1 = 2^{n-1}$, the weight w_0 is arbitrary. It is easy to verify that with this choice of weights, we have

$$F(W,(0,0,...,0)) = w_0; \ F(W,(0,0,...,0,1)) = w_0 + 1; ....;$$

$$..........; \ F(W,(1,1,...,1)) = w_0 + 2^n - 1,$$

i.e., the sequence of weighted sums is a series of consecutive integers starting with w_0. Index the Boolean functions of n variables by different integers from the set $\{0, 1, ..., 2^{2^n} - 1\}$ (the functions may be taken in any order) and define the predicate P_B as follows:

$$P_B(0) = f_0(0), \ P_B(1) = f_0(1), \ ..., \ P(2^n - 1) = f_0(2^n - 1),$$

$$P_B(2^n) = f_1(0), \ ..., \ ..., \ ...,$$

$$P_B(2^n 2^{2^n} - 1) = f_{2^{2^n}-1}(2^n - 1)$$

On the other elements of the ring Z_m, if they exist, the predicate P_B is defined arbitrarily. But last considerations show that every Boolean function of n variables is P-realizable over the ring Z_m with the predicate P_B which has been just built.

Now, considering all $m < 2^{2^n + n}$ and all 2^m possible predicates P_B over the ring Z_m, we can determine the smallest m, for which there exists a predicate P_B that gives a complete weighting description of all Boolean functions of n variables. The upper bound on the cardinality of the residue class ring Z_m admitting a complete weighting description of Boolean functions of n variables obtained in the proof of Theorem 3.3.1 is clearly grossly overestimated. It can be improved by the following considerations.

In the proof of the Theorem 3.3.1, we constructed the sequences $\{P_B\} = \{P_B(0), P_B(1), ..., P_B(m-1)\}$ of values of the predicate P_B on the elements of the ring Z_m such that for any Boolean function there was a subsequence that coincided with sequence of values of this function.

However, we can identify binary sequences shorter than 2^{2^n+n}, which have the following property: all the relevant subsequences occur among the subsequences of length 2^n. We already considered such sequences in the Section 3.1. The smallest length of such a subsequence is 2^{2^n}, and, as we argued in the Section 3.1, they exist for any n. The simplest of such sequences (or a complete code ring for code sequences of a length 2^n), is the complete code ring obtained from maximum-length sequences, or M-sequences that are binary recursive sequence of a special form [Peterson & Weldon (1972)]. Thus, there exists a linear weighting description for all Boolean functions of n variables by $n+1$ weights over the ring Z_m, where $m = 2^{2^n}$. It is important to note that combinations of parameters differing only in the component w_0 also may be taken as descriptions of different functions, in this case. On the basis of this remark, we can again significantly reduce the bound on the cardinality of the residue class ring admitting a complete realization of all Boolean functions of the given number of variables. Indeed, each function of $n+1$ variables is representable as the composition of two functions of n variables. Since both functions have weighting vectors differing only in the weight w_0, the given function of n+1 variables also is P-realizable. Thus a description of all Boolean function of $n+1$ variables exists over the ring Z_m, $m = 2^{2^n}$. Therefore it is clear that all Boolean functions of n variables, $n \geq 1$, are P-realizable over the residue class ring Z_m, $m \leq 2^{2^{n-1}}$.

The upper bound for $n > 2$ is still overestimated, since we obtained it by fixing the weights $w_2, ..., w_n$, which means that we used only m^2 different combinations in weighting vectors $(w_0, w_1, ..., w_n)$ from the total of m^n combinations. Redundancy of this weighting description follows already from the fact that, by Lemmas 3.3.1 and 3.3.2, in order to ensure P-realization of all Boolean functions of n variables, it is necessary and sufficient to obtain a weighting vector for any arbitrary representative of each class of equivalency.

The relationship (3.3.3) leads to a lower bound on the cardinality of the residue class ring Z_m admitting a complete realization of all Boolean functions of n variables: $m \geq 2^{\frac{2^n}{n+1}}$. This bound, in turn, is an underestimate for $n \geq 2$. It was derived relying on the fact that every Boolean function is described over the ring Z_m by precisely one weighting vector. In fact, some Boolean functions may have more than one describing weighting vector. Indeed, consider the weighting vectors of all Boolean functions over the residue class ring Z_m, such that the Predicate P_B takes the value 0 on m_0 elements of the ring and the value 1 on $m_1 = m - m_0$ elements. The Boolean function $f(y_1, \ldots, y_n) \equiv 0$ (the constant 0) is P-realizable in this case by m_0 weighting vectors of the form $(w_0, 0, 0, \ldots, 0)$, where $P(w_0) = 0$. The Boolean function $f(y_1, \ldots, y_n) \equiv 1$ (the constant 1) is correspondingly P-realizable by m_1 weighting vectors of a similar form. Thus, the constants 0 and 1 are P-realizable by m weighting vectors, in which the only nonzero weight is w_0. Now, the constant 0 may be realized by the weighting vector $(w_0, 0, 0, \ldots, 0, w_0' - w_0, 0, \ldots, 0)$, where $P(w_0) = P(w_0') = 0$. Clearly, the total number of different vectors of this form is $nm_0(m_0 - 1)$. The constant 1 correspondingly is realizable by a weighting vector of the same form. Each of the functions $f(y_1, \ldots, y_n) = y_i$ is P-realizable by a weighting vector of the form $(w_0, \underbrace{0, \ldots, 0}_{i-1}, w_1 - w_0, 0, \ldots, 0)$, where $P(w_0) = 0$; $P(w_1) = 1$. The total number of such weighting vectors for each variable function is $m_0 m_1$, and for all variable functions it is $nm_0 m_1$. The same number of weighting vectors describes Boolean functions that are equal to negations of the different variables. We thus obtain the following relationship for the lower bound on m:

$$m^{n+1} - nm(m-1) - m \geq 2^{2^n} - 2n - 2.$$

This result does not constitute a significant change in the order of magnitude of the lower bound on m.

In the process of proving Theorem 3.3.1, we also answered the question of P-realization of all Boolean functions of a given number of variables over some residue class ring Z_m using possibly different predicates P_B. It is easy to show that any function $f(y_1, \ldots, y_n)$ is P-realizable over the residue class ring Z_m with $m = 2^n$ if the values of predicate P_B on the elements of the ring Z_m can be taken as the sequence of values of this function on combinations of values of the variables. Then the function $f(y_1, \ldots, y_n)$ itself is described by the weighting vector $(0, 2^{n-1}, 2^{n-2}, \ldots, 2, 1)$.

A linear weighting parametric description of Boolean functions of n variables over the residue class ring Z_m is called *minimal* if there is no Boolean function of n variables that has such a description over any residue class ring Z_{m_1}, where $m_1 < m$. Recall that for minimal description we have the bound $2^{\frac{2^n}{n+1}} \le m \le 2^{2^{n-1}}$. The lower and upper bounds $B_l^n(m)$ and $B_u^n(m)$ for $n \le 6$ are given in Table 3.3.1.

Table 3.3.1. Estimates for $B_l^n(m)$ and $B_u^n(m)$ for $n \le 6$

n	0	1	2	3	4	5	6
$B_l^n(m)$	2	2	2.52	4	9.19	40.32	565.3
$B_u^n(m)$	-	2	4	16	256	65536	$\approx 4 \cdot 10^9$

A minimal description for Boolean functions of a single variable corresponds to the case where the upper and lower bounds $B_l^n(m)$ and $B_u^n(m)$ are equal. It can be shown that for Boolean functions of two variables there is no complete linear description over the ring Z_3 such that $B_l^n(m) < 3 < B_u^n(m)$ and therefore a minimal description exists only over the ring Z_4, which corresponds to the upper bound $B_u^n(m)$. The predicate P_B that admits a complete realization of all Boolean functions of two variables over the ring Z_4 is not uniquely defined. Just as certain transformations of Boolean realizable functions into realizable functions, so certain operations on predicates can be defined that take realizing predicates into realizing predicates. Let us define the multiplication of the predicate P_B by a ring element $a \in Z_m$ and the addition of the predicate P_B with a ring element $b \in Z_m$ as follows:

$$(aP_B)[z] = P_B[az]; \quad (P_B + b)[z] = P_B[z+b]; \quad a, b, z \in Z_m.$$

Then we can define two operations on realizing predicates, which also produce predicates realizing the same Boolean functions.

Lemma 3.3.6. If the predicate P_B admits a P-realization of some Boolean functions of n variables over the ring Z_m, then the predicate $P + b$, $b \in Z_m$ also admits a P-realization of the same Boolean functions.

Proof. Indeed, if the Boolean function $f(y_1, \ldots, y_n)$ is P-realizable with the weighting vector $(w_0, w_1, \ldots, w_n)$ and with the predicate P_B, then with the predicate $P_B + b$ it will be P-realizable with the weighting vector $(w_0 - b, w_1, \ldots, w_n)$.

Lemma 3.3.7. If the predicate P_B admits a P-realization of some Boolean functions of n variables over the ring Z_m and the numbers a and m

are relatively prime, then the predicate aP_B admits a P-realization of the same Boolean functions.

Proof. If the function $f(y_1, ..., y_n)$ is P-realizable with the weighting vector $(w_0, w_1, ..., w_n)$ and with the predicate P_B, then with the predicate aP_B it is P-realizable with the weighting vector $(a^{-1}w_0, a^{-1}w_1, ..., a^{-1}w_n)$. The existence of the element a^{-1} is ensured by the condition that a and m are relatively prime, and therefore their greatest common divisor is equal to 1.

Let $\Phi(m)$ be the number of elements in the series $0, 1, 2, ..., m-1$ that are relatively prime with m (the Euler function). Then, adding the predicate P_B with elements of the ring Z_m and multiplying the predicate P_B by elements of the ring Z_m relatively prime with m, we can obtain up to $m\Phi(m)$ new predicates that preserve the property of complete P-realization of all Boolean functions of n variables.

Lemma 3.3.8. If the predicate $P_B = P$ is t-odd and the numbers a and m are relatively prime, then the predicate $P' = aP + b$ is t'-odd, where $t' = a^{-1}(t - 2b)$.

Proof. Since a and m are relatively prime, to each $z \in Z_m$ there corresponds a unique element $u \in Z_m$ such that $u = a^{-1}(z - b)$. Then

$$P'[u] = P'[a^{-1}(z-b)] = aP[a^{-1}(z-b)] + b = P[z-b] + b = P[z];$$
$$P'[t' - u] = P'[a^{-1}(t - 2b) - a^{-1}(z-b)] = P'[a^{-1}(t - z - b)] + b =$$
$$aP[a^{-1}(t - z - b)] + b = P[t - z - b] + b = P[t - z].$$

Since the predicate P is t-odd, we have $P[t - z] = 1 - P[z]$, whence $P'[t - u] = 1 - P'[u]$.

Lemma 3.3.9. If the predicate $P_B = P$ admits simultaneous P-realization of the Boolean function $f(y_1, ..., y_n)$ and its negation $\bar{f}(y_1, ..., y_n)$ then the predicate $\overline{P}[z] = 1 - P[z]$, $z \in Z_m$ (negation of predicate P) also admits simultaneous description of these functions.

Proof. Indeed, if with the predicate P the function f is P-realizable by the weighting vector W_1 and the function $\bar{f}$ is P-realizable by the weighting vector W_2, then with the predicate $\overline{P}$ they are P-realizable by the weighting vectors W_2 and W_1, i.e., their weighting vectors are interchanged.

Evidently, that if predicate P ensures P-realization of all Boolean functions of n variables, then the predicate $\overline{P}$ also ensures the same. It is also clear that the negation of a predicate does nor change its t-parity property.

Returning to the problem of minimal linear weighting paramtric description of Boolean functions of two variables over the residue class ring Z_4, we can say that up to invariant operations over the realizing predicates, this description is defined by a unique predicate P_2 whose values on the elements of the ring Z_4 are $P_2[0] = P_2[1] = 0, P_2[2] = P_2[3] = 1$. In the ring Z_m, the set of residue classes whose representatives are relatively prime with m forms a reduced residue system. In this case it consists of two elements (1 and 3). However, the predicate $3P_2$ coincides with the predicate $P_2 + 1$. The predicate $\overline{P_2}$ (the negation of P_2) coincides with the predicate $P_2 + 2$.

Let us consider again the favorite and popular XOR problem. The Table 3.3.2 demonstrates its solution over the ring Z_4 with the weighting vector $(0, 2, 2)$ and the predicate P_2 which was just defined. So this again counter the popular statement that the solution of XOR problem on the single neuron is impossible.

The Table 3.3.3 contains the weighting vectors for all functions of two variables.

Table 3.3.2. Solution of the XOR problem on the single UBN over the ring Z_4

y_1	y_2	$z = w_0 + w_1 y_1 + w_2 y_2$	$P_2(z)$	XOR
0	0	0	0	0
0	1	2	1	1
1	0	2	1	1
1	1	0	0	0

A minimal linear weighting parametric description of Boolean functions of three variables exists over the residue class ring Z_8. Up to invariant operations, it is defined by the unique predicate P_3, for which $P_3[0] = P_3[1] = P_3[2] = P_3[6] = 0, P_3[3] = P_3[4] = P_3[5] = P_3[7] = 1$.

The predicate P_2 is 3-odd, the predicate P_3 is 5-odd, and therefore it is sufficient to find the weighting vectors only for the representatives of each SD-class.

A minimal description of Boolean functions of four variables exists over the residue class ring Z_{18}. One of the realizing predicates, for instance, is the predicate P_4:

$$P_4[0] = P_4[1] = P_4[2] = P_4[3] =$$
$$= P_4[4] = P_4[6] = P_4[8] = P_4[13] = P_4[15] = 0,$$
$$P_4[5] = P_4[7] = P_4[9] = P_4[10] =$$
$$= P_4[11] = P_4[12] = P_4[14] = P_4[16] = P_4[17] = 1.$$

Table 3.3.3. Weighting vectors of Boolean functions of two variables relatively the predicate P_2

Fun #	0	1	2	3	4	5	6	7	8	9	10	11	12	13	14	15
w_0	0	0	0	0	0	0	0	0	2	2	2	2	2	2	2	2
w_1	0	1	3	2	1	0	2	3	3	2	0	1	2	3	1	0
w_2	0	1	1	0	3	2	2	3	3	2	2	3	0	1	1	0

A natural question is how it is possible to find the weighting vectors for the representatives of SD-classes. The following procedure presented in [Aizenberg I. (1985)] is very convenient from the computing point of view.

For given ring Z_m, the predicate P and the function f we have to check, is this function P-realizable, and, if yes, to find at least one weighting vector for it. Let $P^{-1}(s)$, $s \in \{0,1\}$ is the complete prototype of s in the ring Z_m. Then r_j is an arbitrary element of the complete prototype $P^{-1}(f_j)$, $j = 0, 1, ..., 2^n - 1$. The set $P^{-1}(f_j)$ is a permissible set for r_j, and the corresponding values r_j are called permissible. If the function $f(y_1, ..., y_n)$ is P-realizable, it has at least one weighting vector $(w_0, w_1, ..., w_n)$, and $P(w_0 + w_1 y_1 +...+ w_n y_n) \equiv f(y_1, ..., y_n)$ on all domain of the function f. The last equality may be written in the matrix form in following way:

$$P(K_n \tilde{W}) = f,$$

where K_n is the matrix of the conjunctive (Reed-Muller) transform (see (3.3.7)), $\tilde{W} = (w_0, w_n, w_{n-1}, 0, 0, w_{n-2}, 0, ..., ..., ..., 0, w_1, 0, ..., 0)^T$ (in other words the components of vector $\tilde{W}$, which equal to the powers of 2 are equal to the components of further weighting vector in inverted order except w_0), f is a vector of function values. Then we obtain the following:

$$K_n \tilde{W} = (r_0, r_1, ..., r_{2^n-1})^T, \quad \text{or} \quad \text{denoting} \quad R = (r_0, r_1, ..., r_{2^n-1})^T,$$

$K_n \tilde{W} = R$. Let multiply the matrix K_n^{-1} with both sides of the last equality:

$$\widetilde{W} = K_n^{-1} R. \qquad (3.3.10)$$

The system of linear equations (3.3.10) contains $2^n - (n+1)$ homogenous equations for finding the components of vector R (all of them have to be permissible) and $n+1$ equalities for further finding of the non-zero components of the vector $\widetilde{W}$ representing through components of the vector R. The rank of the system of homogenous equations for finding of R is equal to $2^n - (n+1)$. It means that number of free unknowns is equal to $2^n - (2^n - (n+1)) = n+1$. The unknowns $r_0, r_{2^{n-1}}, r_{2^{n-2}}, ..., r_1$ may be always chosen as free because the determinant of matrix which contains coefficients corresponding to other $2^n - (n+1)$ unknowns is always not equal to zero. Basic unknowns are expressed through the free ones, and then attaching the arbitrary permissible values to free unknowns we can find values of basic unknowns. If they will be also permissible then the permissible vector R exists, function f is P-realizable, and its weighting vector should be evaluated from the corresponding equalities of the system (3.3.10). If at least one of the basic unknowns will take not permissible value we have to change the set of values of free unknowns and to repeat the described procedure. If it is impossible to find a permissible vector R, it means that function f is not P-realizable with the given predicate P and residue class ring. The described procedure is a procedure of the directed sorting because it is limited by taking only of the permissible values of $n+1$ free variables.

To illustrate the approach, which has been just described, let us consider example for $n=3$. The system (3.3.10) in matrix form will be the following:

$$
\begin{pmatrix} w_0 \\ w_3 \\ w_2 \\ 0 \\ w_1 \\ 0 \\ 0 \\ 0 \end{pmatrix}
=
\begin{pmatrix}
1 & 0 & 0 & 0 & 0 & 0 & 0 & 0 \\
-1 & 1 & 0 & 0 & 0 & 0 & 0 & 0 \\
-1 & 0 & 1 & 0 & 0 & 0 & 0 & 0 \\
1 & -1 & -1 & 1 & 0 & 0 & 0 & 0 \\
-1 & 0 & 0 & 0 & 1 & 0 & 0 & 0 \\
1 & -1 & 0 & 0 & -1 & 1 & 0 & 0 \\
1 & 0 & -1 & 0 & -1 & 0 & 1 & 0 \\
-1 & 1 & 1 & -1 & 1 & -1 & -1 & 1
\end{pmatrix}
\begin{pmatrix} r_0 \\ r_1 \\ r_2 \\ r_3 \\ r_4 \\ r_5 \\ r_6 \\ r_7 \end{pmatrix}
,
$$

which gives the following system of equations:

$$w_0 = r_0$$

$$w_3 = r_0 + r_1$$

$$w_2 = -r_0 \qquad + r_2$$

$$0 = r_0 - r_1 - r_2 + r_3$$

$$w_1 = -r_0 \qquad \qquad + r_4$$

$$0 = r_0 - r_1 \qquad - r_4 + r_5$$

$$0 = r_0 \qquad - r_2 \qquad - r_4 \qquad + r_6$$

$$0 = -r_0 + r_1 + r_2 - r_3 + r_4 - r_5 - r_6 + r_7.$$

It contains four homogenous equations for finding of R and four equalities for finding of the weighting vector W.

We will return below to the weighting representation of the functions over the residue class ring considering a generalization of the results presented in this Section to the case of multiple-valued functions (see Section 3.5).

4.   *P*-REALIZABLE BOOLEAN FUNCTIONS OVER THE FINITE FIELD

In the Section 3.3 we considered some general features of P-realizable Boolean functions over the algebra A. P-realization of Boolean functions over such finite algebra as residue class ring was considered also at the same section. A special kind of interest is P-realization of Boolean functions over the finite field $GF(2^k)$ and $GF(p^k)$ for any prime p. We will use here the results presented in [Aizenberg I. (1985)] and [Aizenberg & Aizenberg (1991)] with some supplements.

So let the finite field be an algebra A in the Definition 3.1.2.

Theorem 3.4.1. For the given value of n a value of k exists such that it is possible to define a predicate P on the field $GF(2^k)$ that all the Boolean functions of n variables will be P-realizable over the field $GF(2^k)$.

Proof. We will consider the elements of the field $GF(2^k)$ as k-dimensional binary vectors: $a \in GF(2^k)$, $a = (a^{(1)}, a^{(2)}, ..., a^{(k)})$; $a^{(l)} = 0, 1$; $i = 1, .., k$. The addition in the field $GF(2^k)$ is component by component mod 2 addition of binary vectors: $a_1 + a_2 = (a_1^{(1)} \oplus a_2^{(1)}, ..., a_1^{(k)} \oplus a_2^{(k)})$. For the weighting sum of Boolean variables $y_1, ..., y_n$ we obtain the following expression:

$$w_0 + w_1 y_1 + ... + w_n y_n = \left[..., (w_0^{(i)} \oplus w_1^{(i)} y_1 \oplus ... \oplus w_n^{(i)} y_n), ...\right] i = 1, ..., k.$$

Let $w_0 + w_1 \alpha_1 + ... + w_n \alpha_n = z_\alpha$. Then the value of weighted sum corresponding to the values $\alpha_1, ..., \alpha_n$ of the variables is presented by the binary vector $z_\alpha = (z_\alpha^{(1)}, ..., z_\alpha^{(k)})$. Let some Boolean function $f(y_1, ..., y_n) = f_1$ be *P*-realizable over the field $GF(2^n)$. This is always possible because any function of *n* variables is *P*-realizable over the field $GF(2^n)$ with the predicate P_1 whose vector of values on the elements of field $GF(2^n)$ is equal to the vector of values of the function *f*. Let us fix just described predicate P_1 over the field $GF(2^n)$ and create all the possible vectors of values of functions like $f = \left[P_1(z_{\alpha_0}), P_1(z_{\alpha_1}), ..., P_1(z_{\alpha_{2^n-1}}) \right]$. The maximal possible number of functions, which may be obtained by such a way is equal to $N!$, $N = 2^n$, but actually many of them (or even all, if the function f^1 is equal to the constant) will coincide one each other. We will obtain as a result the set Φ_1 of the functions, which are *P*-realizable over the field $GF(2^n)$ with the predicate P_1. Then we will choose some function $f_2 \notin \Phi_1$ and create the class Φ_2 in a similar way. Then we will create the classes $\Phi_3, ..., \Phi_m$. Evidently, such a process is finite because number of classes will be always less then 2^{2^n} (number of all Boolean functions of *n* variables). So all functions of *n* variables are *P*-realizable over the field $GF(2^n)$, but with *m* different predicates $P_1, ..., P_m$. We will show that for any value of *n* it is possible to choose the finite field of such a cardinality, on which it is possible to define a single predicate *P*, which ensures *P*-realization of all functions of *n* variables. Let us consider mod 2 polynomial representation of functions $f_1, ..., f_m$ (they may be obtained for instance, using conjunctive (Reed-Muller) transform, see (3.3.9)) that are generators of the classes $\Phi_1, ..., \Phi_m$. Let us consider the new variables $y_{n+1}, ..., y_{n+m}$, and multiple them with those conjunctions, which have not been presented in corresponding mod 2 polynomial for the given function f_i, $i = 1, ..., m$. We will obtain in such a way *m* new polynomials. Each of them presents the function $F(y_1, ..., y_n, y_{n+1}, ..., y_{n+m})$ of *n+m* variables. It is clear that, if $y_{n+i} = 0$ and $y_{n+1} = y_{n+i-1} = y_{n+i+1} = ... = y_{n+m} = 1$ then $F(y_1, ..., y_n, 1, ..., 1, 0, 1, ..., 1) = f_i(y_1, ..., y_n)$. Evidently, that

function F is P-realizable over the field $GF(2^{n+m})$ with the predicate whose vector of values on the elements of field $GF(2^{n+m})$ is equal to the vector of values of function F. But it means that any function of n variables is P-realizable over the field $GF(2^{n+m})$. Indeed, let $g(y_1, ..., y_n) \in \Phi_i$ and $g = (..., P_i(z_\alpha^{(i_1)}, ..., z_\alpha^{(i_n)}), ...)$. It is evident that the function g is P-realizable over the field $GF(2^{n+m})$ with the weighting vector $(\widetilde{w}_0, \widetilde{w}_1, ..., \widetilde{w}_n)$ whose components are the following:
$\widetilde{w}_0 = (w_0^{(1)}, ..., w_0^{(n)}, 1, ..., 1, 0, 1, ..., 1)$,
$\widetilde{w}_j = (w_j^{(1)}, ..., w_j^{(n)}, 0, ..., 0); j = 1, ..., n$, where $(w_0, w_1, ..., w_n)$ is a weighting vector of the function g obtained for the field $GF(2^n)$ by method just proposed above. Theorem is proven.

The Theorem 3.4.1 may be easily generalized for the finite field $GF(p^k)$ for any prime number p.

Of course, for any value of n it is possible to find the finite field of minimal cardinality over which it is possible to ensure P-realization of all Boolean functions of n variables with the single predicate. For example, all functions of 2 variables are P-realizable over the field $GF(2^3)$, all functions of 3 variables are P-realizable over the field $GF(2^4)$.

Let us return again to our favorite XOR problem. The Table 3.4.1 demonstrates its solution over the field $GF(2^3)$ with the weighting vector $\big((000), (001), (001)\big)$ and the following predicate P:
$$P(000) = P(010) = P(011) = P(101) = 0;$$
$$P(001) = P(100) = P(110) = P(111) = 1.$$

Table 3.4.1. Solution of the XOR problem on the single UBN over the field $GF(2^3)$

y_1	y_2	$z = w_0 + w_1 y_1 + w_2 y_2$	$P_2(z)$	XOR
0	0	000	0	0
0	1	001	1	1
1	0	001	1	1
1	1	000	0	0

Also as in the case of residue class ring considered above (Section 3.3), the search of weighting vectors for all Boolean functions of n variables may be reduced to the solution of such a problem only for the representatives of SD, or PN equivalence classes. A natural question is then to search weighting vectors for these representatives. Such a problem may be reduced to the solution of system of Boolean equations. This idea has been presented in [Aizenberg I. (1984)]. Simultaneously we would like to consider the fast method of solution of Boolean equations proposed in [Aizenberg N. (1991)]. Let us begin from it.

We mean that a Boolean equation is a linear algebraic equation over the field GF(2). Let $\varphi(y_1, \ldots, y_n) = 0$ be a Boolean equation (it is always possible to reduce it to such a form moving all terms to the left side). We will call the order of coefficients of mod 2 polynomial of Boolean function obtained by conjunctive transformation (which is inverse to itself over the field GF(2), see (3.3.7)) the *Hadamard order*, and the corresponding mod 2 polynomial will be called the *Hadamard* mod 2 *polynomial*. Let us reorder the coefficients of the equation $\varphi(y_1, \ldots, y_n) = 0$ to Hadamard order (so in the order corresponding to order of the conjunctions in the matrix of conjunctive (Reed-Muler) transform, see (3.3.7)). Then we have to evaluate the conjunctive spectra (over the field GF(2)) of the vector-signal, in which coefficients of our equation are in the Hadamard order. Evidently, binary numbers, which correspond to the zero-valued coefficients of conjunctive spectra, determine the solutions of our equation.

Example 3.4.1. Let us consider the equation $y_1 + y_1 y_2 + 1 = 0$. Vector of the coefficients in Hadamard order will be the following: $\varphi = (1, 0, 1, 1)^T$. Then

$$
K_2\varphi = \begin{pmatrix} 1 & 0 & 0 & 0 \\ 1 & 1 & 0 & 0 \\ 1 & 0 & 1 & 0 \\ 1 & 1 & 1 & 1 \end{pmatrix}\begin{pmatrix} 1 \\ 0 \\ 1 \\ 1 \end{pmatrix} = \begin{pmatrix} 1 \\ 1 \\ 0 \\ 1 \end{pmatrix} \begin{matrix} 00 \\ 01 \\ 10 \\ 11 \end{matrix} \quad y_1 = 1, y_2 = 0,
$$

so $y_1 = 1$, $y_2 = 0$ is a single solution of our equation.

Let we have the system of s Boolean equations:

$$\varphi_i(y_1, \ldots, y_n) = 0; \ i = 1, \ldots, s. \tag{3.4.1}$$

Let us represent the left parts of all equations in Hadamard form. The binary numbers of zero-valued coefficients of the conjunctive spectra of vector consisted of the coefficients of corresponding equation determine solutions of each equation. Evidently, common solutions for all equations belonging to the system will be the solutions of the system. These common solutions may be obtained using component by component disjunction of the conjunctive spectra of vectors-coefficients of all equations. Evidently, the solutions of the system will be those values of variables on which such a disjunction will be equal to zero. So we just proved the following theorem.

Theorem 3.4.2. Let us have an arbitrary system of Boolean equations like (3.4.1) written in Hadamard form. The binary vectors, which are binary numbers of the zero-valued components of disjunction of conjunctive spectra $K_n \tilde{\varphi}_1 \vee K_n \tilde{\varphi}_2 \vee ... \vee K_n \tilde{\varphi}_s$ (where $\tilde{\varphi}_i$; $i = 1, ..., s$ is a column-vector of coefficients of Hadamard mod 2 polynomial $\varphi_i(y_1, ..., y_n) = 0$), and only these, will be the solutions of our system.

Efficiency of the proposed algorithm of solution of Boolean equations is based on the extremely low computing complexity of the conjunctive transform. It is evident from (3.3.7) that according to the Good theorem [Good (1958)] the matrix K_n may be factorized into matrix product of n equal weakly completed matrixes. For, example, for $n=2$

$$K_2 = \begin{pmatrix} 1 & 0 & 0 & 0 \\ 1 & 1 & 0 & 0 \\ 1 & 0 & 1 & 0 \\ 1 & 1 & 1 & 1 \end{pmatrix} = \begin{pmatrix} 1 & 0 & 0 & 0 \\ 0 & 0 & 1 & 0 \\ 1 & 1 & 0 & 0 \\ 0 & 0 & 1 & 1 \end{pmatrix} \begin{pmatrix} 1 & 0 & 0 & 0 \\ 0 & 0 & 1 & 0 \\ 1 & 1 & 0 & 0 \\ 0 & 0 & 1 & 1 \end{pmatrix}.$$

This means that for performing the conjunctive transformation of order 2^n only $n2^{n-1}$ additions are needed.

Let consider algorithm of the finding the weighting vector of a P-realizable Boolean function of n variables over the field $GF(2^m)$, $n \leq m$. For simplicity we will begin with the example for $n=2$. Let our field is $GF(2^2)$ also for simplicity. Let the predicate be defined in following way: $P(00)=P(01)=P(10)=0$, $P(11)=1$, in other words P may be considered as conjunction of two variables.. $f(y_1, y_2)$ is the function, for which we want to find the weighting vector, $f = (f^{(0)}, f^{(1)}, f^{(2)}, f^{(3)})^T$ is the vector of values of function $f(y_1, y_2)$. Let (w_0, w_1, w_2) be the weighting vector, which we would like to find. Let $w_0 = (u_1, u_2)$; $w_1 = (u_3, u_4)$; $w_2 = (u_5, u_6)$; $u_i \in \{0,1\}$, $i = 1, ..., 6$. For the weighted sum we have the following expressions on all combinations of variables values:

$$z(0, 0) = w_0 = (u_1, u_2);$$
$$z(0, 1) = w_0 + w_2 = (u_1, u_2) + (u_5, u_6);$$
$$z(1, 0) = w_0 + w_1 = (u_1, u_2) + (u_3, u_4);$$
$$z(1, 1) = w_0 + w_1 + w_2 = (u_1, u_2) + (u_3, u_4) + (u_5, u_6).$$

For P-realization of function f the following system of Boolean equations with six unknowns $u_1, ..., u_6$

$$P[z(0,0)] = P[(u_1, u_2)] = f^{(0)};$$
$$P[z(0,1)] = P[(u_1, u_2) + (u_5, u_6)] = f^{(1)};$$
$$P[z(1,0)] = P[(u_1, u_2) + (u_3, u_4)] = f^{(2)};$$
$$P[z(1,1)] = P[(u_1, u_2) + (u_3, u_4) + (u_5, u_6)] = f^{(3)}$$

(3.4.2)

should to have at least one solution. Taking into account that we just defined P as conjunction (see above) we can transform the system (3.4.2) to the following:

$$u_1 u_2 = f^{(0)};$$
$$(u_1 + u_5)(u_2 + u_6) = f^{(1)};$$
$$(u_1 + u_3)(u_2 + u_4) = f^{(2)};$$
$$(u_1 + u_3 + u_5)(u_2 + u_4 + u_6) = f^{(3)}.$$

After transformation of the left parts of equations and their representations to Hadamard mod 2 polynomials we will obtain the system of Boolean equations in Hadamard form. This system should be solved using the transformation K_6 according to the Theorem 3.4.2. If the system has at least one solution then the function f is P-realizable over the field $GF(2^2)$ and has at least one weighting vector. Evidently, if function f is P-realizable over the field $GF(2^2)$ then all its weighting vectors become exhausted by the set of solutions of the system (3.4.2).

Let us consider the case of arbitrary n and m ($n \le m$). Let the predicate P be defined on the field $GF(2^m)$. Its vector of values on the elements of the field $GF(2^m)$ always coincides with the vector of values of some Boolean function of m variables $g(y_1, \ldots, y_m)$. To answer on the question about P-realization of function $f(y_1, \ldots, y_n)$ and to find its weighting vectors (in the case of positive answer) let us introduce the following expressions for the components of the weighting vector:

$$w_0 = (u_0^{(1)}, \ldots, u_0^{(m)});$$
$$w_1 = (u_1^{(1)}, \ldots, u_1^{(m)});$$

(3.4.3)

$$\ldots$$

$$w_n = (u_n^{(1)}, \ldots, u_n^{(m)}).$$

For the weighted sum we have the following expressions on all combinations of the variables values:

$$z(0, \ldots, 0) = w_0 = (u_0^{(1)}, \ldots, u_0^{(m)});$$

$$z(0, \ldots, 0, 1) = w_0 + w_n = (u_0^{(1)}, \ldots, u_0^{(m)}) + (u_n^{(1)}, \ldots, u_n^{(m)});$$

$$\ldots$$

$$z(1, \ldots, 1) = w_0 + w_1 + \ldots + w_n = (u_0^{(1)}, \ldots, u_0^{(m)}) + \ldots + (u_n^{(1)}, \ldots, u_n^{(m)}).$$

Assuming that the function $f(y_1, \ldots, y_n)$ is P-realizable we obtain the following system, which is similar to system (3.4.2):

$$P[z(0,\ldots,0)] = P[(u_0^{(1)}, \ldots, u_0^{(m)})] = f^{(0)};$$

$$P[z(0,\ldots,0,1)] = P[(u_0^{(1)}, \ldots, u_0^{(m)}) + (u_n^{(1)}, \ldots, u_n^{(m)})] = f^{(1)}; \tag{3.4.4}$$

$$\ldots$$

$$P[z(1,\ldots,1)] = P[(u_0^{(1)}, \ldots, u_0^{(m)}) + \ldots + (u_n^{(1)}, \ldots, u_n^{(m)})] = f^{(2^n-1)}.$$

The system (3.4.4) is a system of 2^n Boolean equations with $m(n+1)$ unknowns $u_0^{(1)}, \ldots, u_0^{(m)}, u_1^{(1)}, \ldots, \ldots, \ldots, u_n^{(1)}, \ldots, u_n^{(m)}$. Taking into account that P is considered as Boolean function $g(y_1, \ldots, y_m)$ of m variables, and reducing all the equations to Hadamard form, we will obtain the system of Boolean equations like (3.4.1). This system should be solved using the algorithm described above. If the system has at least one solution then the function $f(y_1, \ldots, y_n)$ is P-realizable. Moreover, if function $f(y_1, \ldots, y_n)$ is P-realizable then all its weighting vectors become exhausted by the set of solutions of the system (3.4.4).

The Table 3.4.2 contains the weighting vectors of all Boolean functions of two variables over the field $GF(2^3)$ obtained using the algorithm just described.

The algorithm may be generalized for the field $GF(p^m)$ for arbitrary prime p. This generalization is based on the following approach to solution of the equations over the field $GF(p)$. We will use the following transform instead of conjunctive transform:

$$M_p = \begin{pmatrix} 1 & 0 & 0 & \ldots & 0 \\ 1 & 1 & 1^2 & \ldots & \cdot \\ 1 & 2 & 2^2 & \ldots & \cdot \\ \cdot & \cdot & \cdot & \ldots & \cdot \\ 1 & p-1 & (p-1)^2 & \ldots & (p-1)^{p-1} \end{pmatrix}. \tag{3.4.5}$$

Table 3.4.2. Weighting vectors of Boolean functions of two variables relatively the predicate:
$$P(000) = P(010) = P(011) = P(101) = 0;$$
$$P(001) = P(100) = P(110) = P(111) = 1.$$
The table contains only one weighting vector per function

fun #	0	1	2	3	4	5	6	7	8	9	10	11	12	13	14	15
w_0	0	0	0	0	0	0	0	0	1	0	0	1	0	1	1	0
	0	0	0	0	0	0	0	0	0	0	0	0	0	0	0	0
	0	0	0	0	0	0	0	0	0	1	1	0	1	0	0	1
w_1	0	0	1	0	1	0	0	0	0	0	0	1	0	1	0	0
	0	1	1	0	0	0	0	0	0	0	0	0	0	1	1	0
	0	0	0	1	1	0	1	1	1	1	0	1	1	0	0	0
w_2	0	1	1	0	1	0	0	1	1	0	0	1	0	1	1	0
	0	0	0	0	1	0	0	1	0	0	0	1	0	0	0	0
	0	1	1	0	0	1	1	0	1	1	1	0	0	1	1	0

For example, for $p=3$ $M_3 = \begin{pmatrix} 1 & 0 & 0 \\ 1 & 1 & 1 \\ 1 & 2 & 1 \end{pmatrix}$. Kronecker n^{th} power of matrix M_p defines the matrix $\tilde{M}_p$ whose columns satisfy the following equality

$$\tilde{M}_p(\alpha_1, ..., \alpha_n) = y_1^{\alpha_1} ... y_n^{\alpha_n}$$ where

$(\alpha_1, ..., \alpha_n), \alpha_i \in \{0, 1, ..., p\}, i = 0, 1, ..., n$ is the number of the corresponding column. It should be noted that the matrix K_n of conjunctive transform is built according to the same law with $p=1$. For instance, for $p=3$ the matrix $\tilde{M}_3 = M_3^{\otimes 2}$ is

$$\tilde{M}_3 = \begin{pmatrix} 1 & 0 & 0 & 0 & 0 & 0 & 0 & 0 & 0 \\ 1 & 1 & 1 & 0 & 0 & 0 & 0 & 0 & 0 \\ 1 & 2 & 1 & 0 & 0 & 0 & 0 & 0 & 0 \\ 1 & 0 & 0 & 1 & 0 & 0 & 1 & 0 & 0 \\ 1 & 1 & 1 & 1 & 1 & 1 & 1 & 1 & 1 \\ 1 & 2 & 1 & 1 & 2 & 1 & 1 & 2 & 1 \\ 1 & 0 & 0 & 2 & 0 & 0 & 1 & 0 & 0 \\ 1 & 1 & 1 & 2 & 2 & 2 & 1 & 1 & 1 \\ 1 & 2 & 1 & 2 & 1 & 2 & 1 & 2 & 1 \end{pmatrix}.$$

Taking into account that $\tilde{M}_p$ is the Kronecker n^{th} power of matrix M_p, according to [Good (1958)] it may be factorized into the matrix product of n

weakly completed matrices, and therefore the transform defined by matrix $\widetilde{M}_p$ has the fast algorithm. Let us have p-ary equation with n unknowns written in Hadamard form. Evidently, zero-valued spectral coefficients of $\widetilde{M}_p$ - spectrum of vector of the equation coefficients define the solutions of this equation. For solving the system of p-ary equations it is necessary to find $\widetilde{M}_p$ - spectra for vectors of coefficients of all equations written in Hadamard form and then to perform the operation of component by component evaluation of maximums over all the spectra. Zero-valued components of the resulting vector defined the solutions of system by their p-mary numbers.

In this way the computation of all the weighting vectors for P-realizable Boolean function over the field $GF(p^m)$ is reduced to the creation of the system of p-ary equations like (3.4.4), followed by its solution applying the transformation $\widetilde{M}_p = M_p^{\otimes n}$ using the method just described. It means that the algorithm of finding of the weighing vectors over the field $GF(p^m)$ for given function is analogous to the algorithm for the field $GF(2^m)$, which has been described above.

It should be also noted that some of the weighting vectors for P-realizable Boolean function over the finite field may be found applying the algorithm proposed in the Section 3.3 for residue class rings. Indeed, all considerations connected with that algorithm may be repeated for the finite field by simple replacement of the Z_m with $GF(p^m)$. The advantage of the algorithms described in this Section specially for the finite fields, is that they not only give answer on the question about P-realization of the partial function over the partial field. In the case of positive answer the described algorithms make possible to find all the weighting vectors for the corresponding function.

5. POLYNOMIAL AND LINEAR WEIGHTING DESCRIPTIONS OF K-VALUED FUNCTIONS OVER THE ALGEBRA A

Successful generalization of the notion of P-realizable Boolean function on the case of arbitrary algebra and introduction of the notion of PW-realizable Boolean function (Definition 3.3.1), consideration of the P-realizable Boolean functions not only over the field of the complex numbers, but over residue class ring and finite field makes natural to consider the same generalization for multiple-valued (k-valued) functions. Multiple-valued functions over the field of the complex numbers are considered in details above (Chapter 2). The direct generalization of the basic idea [Aizenberg N. et al. (1971ab, 1973)] of approach presented here (Chapter 2) on the case of finite field has been done in [Labunets & Sitnikov (1975)].

It will be natural to make the analogous general remark concerning k-valued threshold functions over the algebra A (as was made for Boolean functions in the Section 3.3.), and to consider the k-valued threshold functions over the residue class ring. We will use here the results of paper [Aizenberg N. et al. (1989)] with some supplements.

Let A be an algebra with a subset of elements $E_k = \{0, 1, ..., k-1\}$; the properties of the elements of the set E_k are not specified at this point. We consider the functions $f(\alpha_1, ..., \alpha_n)$ whose arguments are defined on the set E_k and such that $f(\alpha_1, ..., \alpha_n) \in E_k$ when all $\alpha_i \in E_k$, $i = 1, ..., n$. Let $W = (w_0, w_1, ..., w_l)$ be a vector whose components belong to A, $Y = (y_0, y_1, ..., y_n)$ be a vector with components from E_k. Let $F(W, X)$ be a function defined on the combination $W \in A^l$ and $Y \in E_k^n$, where A^l is the direct l^{th} power of the set of elements A. Let E_k^n be a direct n^{th} power of the set E_k taking values in A. Let P be some k-valued predicate defined on the elements of the algebra A. We say that the k-valued logic function $f(Y) = f(y_1, ..., y_n)$ is described over the given algebra A by the weighting vector W with given F and P (in other words $f(y_1, ..., y_n)$ is a PW- k-valued threshold function), if for each combination of the argument values $\alpha = (\alpha_1, ..., \alpha_n)$; $\alpha_i \in E_k$, $i = 1, ..., n$ the following equality is true:

$$P[F(W,Y)] = f(y_1, ..., y_n). \qquad (3.5.1)$$

We have a complete implementation of k-valued functions $f(X)$ of a given number of variables over the given algebra A with some F and P, if for each k-valued function there exists a weighting vector describing this function. As for weighting parametric descriptions of Boolean functions (see top of the Section 3.3), we can use the conventions regarding the operations "+" and "•" in the algebra A, and the meaning of the elements 0 and 1. We can also assume that the function F of the weighting vector W is a polynomial of degree kn in the variables $y_1, ..., y_n$, i.e., we may consider weighting functions F of the form

$$F(W,\,Y) = v_0 + v_1 y_n + v_2 y_n^2 + v_3 y_n^3 + \ldots +$$

$$+ \ldots + v_{k-1} y_n^{k-1} + v_k y_{n-1} + v_{k+1} y_{n-1} y_n + v_{k+2} y_{n-1} y_n^2 + \ldots +$$

$$+ \ldots\ldots\ldots\ldots\ldots\ldots\ldots\ldots\ldots\ldots\ldots\ldots\ldots\ldots\ldots\ldots\ldots\ldots\ldots + \tag{3.5.2}$$

$$+ v_{k^n-1} y_1^{k-1} y_2^{k-1} \ldots y_n^{k-1},$$

where as before $l+1$ coefficients may take nonzero values, $v_{i_0} = w_0$, $v_{i_1} = w_1$, ..., $v_{i_l} = w_l$, and the remaining $k^n - l - 1$ coefficients take only zero values. This description of k-valued logic functions follows from the equation

$$P[\tilde{M}_k V] = f, \tag{3.5.3}$$

which is an analog of the equation (3.3.8) for the polynomial description of Boolean functions. In the equation (3.5.3) V is the column of coefficients of the weighting polynomial (3.5.2); the matrix $\tilde{M}_k$ is an analog of the conjunctive matrix: it is the Kronecker n^{th} power of the Vandermonde matrix $\tilde{M}_k = (M_k)^{\otimes n}$ (see (3.4.5), where the same matrix has been introduced)

$$M_k = \begin{pmatrix} 1 & 0 & 0 & \ldots & 0 \\ 1 & 1 & 1^2 & \ldots & 1 \\ 1 & 2 & 2^2 & \ldots & 2^{k-1} \\ . & . & . & \ldots & . \\ 1 & k-1 & (k-1)^2 & \ldots & (k-1)^{k-1} \end{pmatrix};$$

$f = (f^{(0)}, f^{(1)}, \ldots, f^{(k^n-1)})^T$. If the Vandermonde determinant over the algebra A does not vanish, i.e., if the matrix $\tilde{M}_k$ is invertible over the algebra A, then we can define with a fixed k-valued predicate P for each k-valued function its realizing polynomial F whose coefficients satisfy the condition

$$V = \tilde{M}_k^{-1} P^{-1}[f],$$

which is an analog of the corresponding condition for Boolean functions (see (3.3.8)). However, with this choice of weighting parametric description of k-valued logic functions, there is no direct analogy between descriptions of Boolean and k-valued functions. In particular, it is well-known that Boolean functions are representable by mod 2 polynomials, which corresponds to a description by 2^n parameters over the residue class ring Z_2 (finite field GF(2)). At the same time, k-valued functions with nonprime k are not representable as polynomials mod k, i.e., they do not have a parametric description with k^n parameters and weighting function (3.5.2) over the ring

Z_k. This follows, in particular, from the fact that the matrix $\tilde{M}_k$ is not in general invertible over Z_k.

We speak about linear weighting parametric description of k-valued logic functions, if the function $F(W, Y)$ has the form

$$F(W,Y) = w_0 + w_1 y_1 + ... + w_n y_n .$$

In this case, there exist some invariant transformation that preserves the realizability of k-valued functions. E.g., permutation of arguments preserves a realizability. We define the negation of an argument of a logic function in the form $\bar{y} = 1 - y$. Then the transformation of a k-valued function involves the negation of some argument also preserves the realizability of the logic function. Let us decompose the function $f(y_1, ..., y_n)$ by the variable y_1:

$$f(y_1, ..., y_n) = \bigvee_{\sigma \in E_k} f_\sigma(y_1) \& f_\sigma(y_2, ..., y_n)$$

where $\quad f_\sigma(y_2, ..., y_n) = f(\sigma, y_2, ..., y_n), \quad f_\sigma(y) = \begin{cases} k-1, & \sigma = y \\ 0, & \sigma \neq y \end{cases}$,

operations $\vee$ and $\&$ (analogues of conjunction and disjunction), respectively are defined as $y_1 \& y_2 = \min(y_1, y_2)$, $y_1 \vee y_2 = \max(y_1, y_2)$. Then if the function $f(y_1, ..., y_n)$ is k-valued threshold function and has a weighting vector, the functions f_σ, $\sigma = 0, 1, ..., k-1$ also are k-valued threshold functions, and among the weighting vectors for the functions f_{σ_1} and f_{σ_2} there are always descriptions that differ only by the weight w_0, and moreover there is always a set of weighting vectors for the functions f_σ with equal weights $w_1, ..., w_n$ such that that the weighting vectors for the functions f_σ and $f_{\sigma+1}$ satisfy the equality $w_0^{(\sigma+1)} - w_0^{(\sigma)} = const$, $\sigma = 0, 1, ..., k-2$. Conversely, if for k functions $f_0, f_1, ..., f_{k+1}$ of n-1 variables there are descriptions with equal weights $w_1, ..., w_n$ and weights $w_0^{(\sigma)}$ satisfying the above requirement, then the k-valued logic function of n variables formed as a composition of the given functions of n-1 variables also is a threshold function over the algebra A.

Let the predicate P be t-odd (see for Boolean functions the Section 3.3). We can show that for appropriate definition of dual function and of a function dual with respect to some variable, transformations passing these dual functions with a t-odd predicate P also preserve realizability property.

Let $A = Z_m$, i.e., algebra A is the residue class ring mod m, and $m \geq k$. We will consider the weighting parametric description of k-valued logic functions which differs from the one considered above. Take a system of k k-valued logic functions of a single variable in the following form:

$f_0 = 1$ - the constant 1;

$f_1 = y$ - a variable;

$f_2 = y(y-1)/2$, etc. ... until the function

$$f_{k-1} = \prod_{i=0}^{k-1} (y-i)/(k-1)!\ .$$

Note that over the ring Z_m for $m \geq k$, all the functions $f_0, f_1, ..., f_{k-1}$ are defined, but we can not use distributivity of multiplication in order to evaluate the functions $f_2, ..., f_{k-1}$, i.e., for instance, $F_2 = y(y-1)/2 \neq y^2/2 - y/2$, because the value of ½ is not necessarily defined in the ring Z_m. From the matrix L_k, entering the values of the function f_i in column i. This gives the matrix

$$L_k = \begin{pmatrix} 1 & 0 & 0 & 0 & \cdots & 0 \\ 1 & 1 & 0 & 0 & \cdots & 0 \\ 1 & 2 & 1 & 0 & \cdots & 0 \\ 1 & 3 & 3 & 1 & \cdots & 0 \\ \cdot & \cdot & \cdot & & \cdots & \cdots \\ 1 & k-1 & \dfrac{(k-1)(k-2)}{2} & \dfrac{(k-1)(k-2)(k-3)}{6} & \cdots & \dfrac{\prod_{i=1}^{k}(k-i)}{(k-1)!} \end{pmatrix},$$

i.e., a Pascal matrix of order k [Aizenberg N. et al. (1984)], whose elements are the binomial coefficients: $L_k = \|l_{ij}\|$, $l_{ij} = C_i^j$, $i \geq j$; $l_{ij} = 0$, $i < j$. If $m \geq k$, the matrix L_k is invertible, and $L_k^{-1} = \|l_{ij}^{(-1)}\|$, where $l_{ij}^{(-1)} = (-1)^{i+j} l_{ij}$. The functions f_i, $i = 0, 1, ..., k-1$, will be called Pascal quasi-polynomials of a degree i. Consider the weighting parametric description of k-valued logic functions over the residue class ring Z_m ($m \geq k$) of the form

$$F(W,\ Y) = v_0 + v_1 f_1(y_n) + v_2 f_2(y_n) + ... +$$
$$+ ... + v_{k-1} f_{k-1}(y_n) + v_k f_1(y_{n-1}) + v_{k+1} f_1(y_{n-1}) f_1(y_n) +$$
$$+ v_{k+2} f_1(y_{n-1}) f_2(y_n) + ... + v_{\alpha_1 k^{n-1} + \alpha_2 k^{n-2} + ... + \alpha_{n-1}k + \alpha_n} f_{\alpha_1}(y_1) ... f_{\alpha_n}(y_n) + ... + \qquad (3.5.4)$$
$$+ ... + v_{k^n - 1} f_{k-1}(y_1) ... f_{k-1}(y_n);\ \ \alpha_i \in E_k,\ i = 1, ..., n.$$

We call it quasi-polynomial weighting parametric description, or Pascal description, for the functions of the k-valued logic of n variables. The main equation in this case is

$$P[L_k^n V] = f\ ,$$

where $I_k^n = (L_k)^{\otimes n}$. Note, in particular, that a quasipolynomial description of k-valued logic functions of n variables by k^n parameters always exists over the ring k. Using the representation (3.5.4), we can define a criterion of represantability of a k-valued function over Z_k for nonprime k in the form of an ordinary polynomial.

Consider the linear weighting parametric description of k-valued functions over the residue class ring Z_m. Repeating the same argument as for Boolean functions, we can show that for any number of variables n there is $m=m(n, k)$ such that over the residue class ring Z_m there exists a complete linear weighting parametric description (so exactly description with $n+1$ weights) of k-valued logic functions of n variables. For m we obtain the following bound estimate:

$$k^{\frac{k^n}{n+1}} \leq m \leq k^n k^{k^n}.$$

By the last considerations we complete the observation of general approach to weighting description of multiple-valued functions and their implementation especially over the residue class ring.

6.   CONCLUSIONS

The theory P-realizable Boolean functions has been developed in this Chapter. On the base of this theory the universal binary neuron (UBN) has been introduced as a neural element, which performs an input/output mapping described by P-realizable Boolean function. The following points are the key aspects for the presented approach: 1) the activation function of the MVN is a function of the argument of the weighted sum; 2) The weights are complex numbers. The features P-realizable Boolean functions are studied. The notion of P-realizable Boolean function has been generalized for the residue class rings and the finite field. Some methods of UBN synthesis are considered. Classification of the Boolean functions in respect to the P-realizability feature has been considered. The learning algorithm for UBN with two linear learning rules will be considered below in the Chapter 4, and applications of UBN-based cellular neural network to solving of image processing problems will be considered in the Chapters 5-6.

Chapter 4

Learning Algorithms

The learning algorithms for multi-valued and universal binary neurons will be considered in this chapter. It will be shown that learning of MVN and UBN should be based on the same principles that perceptron learning. A key principle is to correct the weights with the aim to implement a given mapping between inputs and output of a neuron. It will be shown that the learning for MVN is connected with the notion of k-edge (see Section 2.3). The notion of k-separation of n-dimensional space will be also presented. Two linear correction rules for the implementation of learning algorithm will be considered. A convergence of learning algorithm with both rules will be proven. It will be shown that the learning of UBN may be reduced to the learning of MVN. At the same time the separate learning algorithm for UBN will be considered.

1. LEARNING ALGORITHM FOR MULTI-VALUED NEURON

We will not consider here the learning theory in general. The reader who are interested may find a nice observation of the subject for example, in [Haykin (1994)], [Hassoun (1995)], and many other books and papers. Our subject is the learning of MVN and UBN.

Since it is possible to consider the MVN as a specific generalization of the threshold neural element, it is natural to generalize the principles of perceptron learning to the case of MVN. Indeed, if k=2 in (2.2.1) and (2.2.2), then the equation (2.2.2) transforms to the equation (3.1.4), and MVN becomes a complex-threshold element [Aizenberg N. & Ivaskiv (1977)]. Such an element implements the complex-threshold Boolean functions (see Definition 3.1.1). But according to the Theorem 3.1.3 the set of Boolean complex-threshold functions coincides with the set of Boolean threshold functions. It means that the complex-threshold neural element implements the same threshold Boolean functions as the threshold neural element.

The learning of a threshold element may be presented in the following way. Let us have a learning set $A = A_1 \bigcup A_{-1}$, where A_1 is a subset of the neuron's inputs, on which output has to be equal to 1, A_{-1} is a subset of the neuron's inputs, on which output has to be equal to -1. Learning is reduced to the search of a hyperplane, which separates the subsets of the learning set in the space. The coefficients of a hyperplane equation will be the synaptic weights implementing a corresponding mapping.

Taking into account that MVN implements a k-valued mapping, it is easy to conclude that the learning set of MVN consists of k classes. Let $k>2$ be some integer. Let us consider $(n+1)$ - dimensional vectors $X = (1, x_1,...,x_n)$, $x_1,...,x_n \in \{\varepsilon^0, \varepsilon,..., \varepsilon^{k-1}\}$, where $\varepsilon = i2\pi / k$ is a primitive k^{th} root of unity. Let A_j be a learning subset $\{X_1^{(j)},... X_{N_j}^{(j)}\}$ of the input neuron states corresponding to the output value j (ε^j in the complex-valued form). In such a case we can present the global learning set as $A = \bigcup_{0 \le j \le k-1} A_j$. Certainly, some of the sets A_j may be empty, and $A_i \bigcap A_j = \varnothing$ for any $i \ne j$.

Definition 4.1.1. The sets $A_0, A_1,..., A_{k-1}$ are called k-*separated*, if it is possible to find a permutation $R = (\alpha_0, \alpha_1, ..., \alpha_{k-1})$ of the elements of the set $E_k = \{0, 1, ..., k-1\}$, and a weighting vector $W = (w_0, w_1,..., w_n)$ such that

$$P(X, \overline{W}) = \varepsilon^{\alpha_j} \qquad (4.1.1)$$

for each $X \in A_j, j = 0, 1, ..., k-1$. Here $\overline{W}$ is a complex-conjugated to W vector, $(X, \overline{W})$ is a scalar product of the $(n+1)$-dimensional vectors within the $(n+1)$-dimensional unitary space, P is the MVN activation function (2.2.2).

It is clear that the learning problem for the MVN may be reduced to the problem of k-separation of learning subsets. In other words, the learning problem for the given learning subsets $A_0, A_1, ..., A_{k-1}$ is a problem, how to find a permutation $(\alpha_0, \alpha_1, ..., \alpha_k)$ and a weighting vector $W = (w_0, w_1, ..., w_n)$, which satisfy the condition (4.1.1).

The notion of k-separation is closely connected with the notion of edge-like sequence (see Definition 2.3.6). Evidently, if the sets $A_0, A_1,..., A_{k-1}$

are *k*-separated, and the permutation $R = (\alpha_0, \alpha_1, ..., \alpha_{\kappa-1})$ will be applied to them then the edge-like sequence will be result of such a permutation. From the geometrical point of view the *k*-separation means that elements from only the class $A_j = \{x | f(x) = \varepsilon^j\}$ belong to each edge of the *k*-edge. Moreover, the elements belonging to the same class could not belong to the different edges.

The learning sequence from the set *A* is any infinite sequence S_u of objects $u_0, u_1, ..., u_{k-1}$ such that $u_m \in S_u \Rightarrow u_m \in A$, $u \in A \Rightarrow u = u_m$, where *m* takes values from the infinite set.

The learning is reduced to obtaining the sequence S_w of the weighting vectors $W_0, W_1, ...$. Each vector corresponds to single iteration of learning. We say that the learning process has converged if beginning from the some number m_0: $W_{m_0} = W_{m_0+1} = ...$, and (4.1.1) also as (2.2.3) are true. We will consider here two learning rules, but first of all let us consider a global scheme for learning. Two strategies may be used. It will be shown below that they are equivalent, moreover, other equivalent strategies may be proposed.

Strategy 1 (Sequential sorting of the learning subsets).

Step 1. The starting vector W_0 for the weights is chosen as an arbitrary vector (e.g., its components may be the random numbers or may be equal to zero); *j*=0;

Step 2. Checking of the (4.1.1) for the elements of learning subsets $A_0, ..., A_j$:

> *if* (4.1.1) is true for all of them
> *then go to* the step 4
> *else go to* the step 3;

Step 3. Obtaining of the vector W_{m+1} from the vector W_m by the learning rule (considered below);
> *go to* the step 2;

Step 4. $j = j+1$; *if* $j = k$ *then* learning is finished successfully, *else go to* the step 2.

Strategy 2. (Consideration of all learning subsets together).

X_s^j is the sth element of the learning set *A* belonging to the learning subset A_j. Let *N* be the cardinality of the set *A*.

Step 1. The starting weighting vector W_0 is chosen as an arbitrary vector (e.g., its components may be the random numbers or may be equal to zero); *s*=1; *l*=0;

Step 2. Checking of the (4.1.1) for the element $X_s^{j_*}$ of learning subset

A_{j_*}:

    if (4.1.1) is true for it
    then go to the step 4
    else begin l=1; *go to* the step 3 *end*;

Step 3. Obtaining of the vector W_{m+1} from the vector W_m by the learning
      rule (considered below);

    go to the step 2;

Step 4. $s = s+1$;   *if* $s \le N$
          then go to the step 2
          else if l=0
              then the learning is finished successfully
              else begin s=1; l=0; *go to* the step 2; *end*.

l is a flag of the successful learning for some vector X_s^j in Strategy 2. If
current vector W satisfies (4.1.1) for the vectors from learning set then l=0; if
the vector W does not satisfy (4.1.1) for some vector X_s^j then l=1.

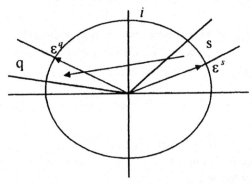

Figure 4.1.1 Problem of the MVN learning

    A learning rule is of course a key point of the learning algorithm with
both strategies just proposed. It should ensure an effective correction of the
weights, and lead to convergence as soon as possible. We will consider here
two linear learning rules. Both of them may be considered as generalization
of the corresponding perceptron learning rules. A generalization means the
following. If perceptron output for some element of the learning set is
incorrect (1 instead of -1, or -1 instead of 1) then the weights should be
corrected by some rule to invert a sign of the weighted sum. Therefore, it is
necessary to move the weighted sum to the opposite subdomain (respectively
from "positive" to "negative", or from "negative" to "positive"). For MVN,
which performs a mapping described by k-valued function, we have exactly
k domains. Geometrically they are the sectors on the complex plane. If the

desired output of MVN on some vector from the learning set is equal to ε^q then the weighted sum should to fall into the sector number q. But if the actual output is equal to ε^s then the weighted sum has fallen into sector number s (see Fig. 4.1.1). A learning rule should correct the weights to move the weighted sum from the sector number s to the sector number q.

2. THE FIRST LEARNING RULE FOR THE MVN LEARNING ALGORITHM

One of the simplest learning rules for perceptron (see e.g., [Haykin (1994)] is the following:

$$W_{m+1} = W_m + (-1)X,$$

where W_m and W_{m+1} are respectively, current and following weighting vectors, X is a current element (vector) of the learning set, -1 is a natural correction coefficient (indeed, if the current value of the weighted sum is incorrect, it should be "moved" to change its sign).

The following rule has been proposed in [Aizenberg N. & Ivaskiv (1977)] and developed in [Aizenberg & Aizenberg (1992), (1993)] to correct a weighting vector in MVN learning algorithm (to obtain the vector W_{m+1} from the vector W_m):

$$W_{m+1} = W_m + \omega_m C_m \varepsilon^\alpha \overline{X}, \tag{4.2.1}$$

where $\overline{X}$ is the complex-conjugate of the neuron's input signals (the current vector from the learning set), ε^α is a desired neuron's output (in complex form, look (2.2.1) for the correspondence), C_m is a scale coefficient, ω is a correction coefficient. Such a coefficient must be chosen from the point of view that the weighted sum should move exactly to the desired sector or at least as close as possible to it, after correction of the weights according to the rule (4.2.1). For the choice of ω the following three cases should be considered. Let $\varepsilon^\beta = (X, \overline{W}_m)$ be an actual neuron's output, i is imaginary unity.

Case 1. $\omega = -i\varepsilon$, if $\varepsilon^\beta = \varepsilon^{\alpha+1}$ for $k=2$ and $k=3$, or $\varepsilon^{\alpha+1} \prec \varepsilon^\beta \prec \varepsilon^{\alpha+[k/4]}$ for $k \geq 4$.

Case 2. $\omega = 1$, if $\varepsilon^{\alpha+[k/4]+1} \prec \varepsilon^\beta \prec \varepsilon^{\alpha+3[k/4]-1}$ for $k \geq 4$ (for $k<4$ such a case is impossible).

Case 3. $\omega = i$, if $\varepsilon^\beta \prec \varepsilon^{\alpha+2}$ for $k=3$, or $\varepsilon^{\alpha+3[k/4]} \prec \varepsilon^\beta \prec \varepsilon^{\alpha+k-1}$ for $k \geq 4$ (for $k=2$ such a case is impossible).

Here [$k/4$] is an integer part of $k/4$ and a relation noted by $\prec$ is defined in the following way: $\varepsilon^p \prec \varepsilon^q \Leftrightarrow p(\mathrm{mod}\, k) \leq q(\mathrm{mod}\, k)$.

Case 1 corresponds to a motion to the "right" side from the actual sector, if difference between arguments of the actual and desired output is not more than $\pi/2$. Case 3 corresponds to a motion to the "left" side from the actual sector, if difference between arguments of the actual and desired output is not more than $\pi/2$. Case 2 corresponds to the situation, where the difference between arguments of the actual and desired output is more than $\pi/2$.

A role of the scale coefficient C_m in the rule (4.2.1) should be clarified. This coefficient may be changed during the learning process, and changing C_m it is possible to control the speed of the learning process (to accelerate it). We will return to the role of this coefficient in the Section 4.5. Anywhere C_m should contain the multiplier $\dfrac{1}{(n+1)}$ because the weighting vector W is exactly (n+1)-dimensional vector. Without such a multiplier the modulo of the second (correcting) addend in the equality (4.2.1) will be (n+1)-times higher than necessary. Indeed, let obtain the weighted sum after correction according to the rule (4.2.1):

$$(X, \overline{W}_{m+1}) = (X, \overline{W}_m) + C_m(X, \omega_m \varepsilon^\alpha \overline{X}) =$$
$$(X, \overline{W}_m) + C_m \omega_m \varepsilon^\alpha (X, X) = (X, \overline{W}_m) + (n+1)C_m \omega_m \varepsilon^\alpha.$$

Theorem 4.2.1. (About the convergence of the learning algorithm with the learning rule (4.2.1)). Let the learning subsets $A_0, A_1, \ldots, A_{k-1}$ are k-separated according to the Definition 4.1.1. Then such an integer number m_0 ($1 \leq m_0 \leq k!$) exists that $W_{m_0+1} = W_{m_0+2} = \ldots = W_{m_0+3} = \ldots$, where $W_{m_0+1} = W$, and permutation $R = R_{m_0}$ satisfy the equation (4.1.1). In other words: If the learning subsets $A_0, A_1, \ldots, A_{k-1}$ are k-separated for the given value of k, then the learning algorithm with the rule (4.2.1) converges, and requests for convergence not more than $k!$ iterations (the extreme upper estimate).

Proof. Suppose that conclusion of the theorem is false. It means that we can't get a convergence of the sequence S_w of the weighting vectors. Therefore a correction with the rule (4.2.1) gives infinite number of the new weighting vectors, which do not satisfy the condition (4.1.1) at least for one element from some learning subset. According to our assumption the learning subsets $A_0, A_1, \ldots, A_{k-1}$ are k-separated. Therefore the weighting vector W and permutation $R = (\alpha_0, \alpha_1, \ldots, \alpha_{k-1})$, which are satisfied

(4.1.1), exist. Let us consider only the learning subsets $A'_j = \varepsilon^{-\alpha_j} A_j = \left\{ \varepsilon^{-\alpha_j} X_1^{(j)}, \ldots, \varepsilon^{-\alpha_j} X_{N_j}^{(j)} \right\}$, $j = 0, 1, \ldots, k\text{-}1$ for simplification. The set $A' = \bigcup_{0 \le j \le k-1} A'_j$ is the reduced learning set. We will say that this set is separable, if $A_0, A_1, \ldots, A_{k-1}$ are k-separated. Thus, if A' is separable, it means that such a vector W exists that $P(X', \overline{W}) = \varepsilon^0$ for all $X' \in A'$. The last condition is true if and only if (4.1.1) is true.

Let $S_{X'} = X'_1, X'_2, \ldots$ be a learning sequence with the elements from the reduced learning set A'. It is evident that $X'_m = \varepsilon^{-\alpha_j} X_m$, if $X_m \in A_j$, $j = 0$, 1, ..., k-1; $m = 0, 1, 2, \ldots$. Let us suppose $C_m = 1$ in equation (4.2.1) without loss of generality. We can transform the equation (4.2.1) taking into account the last considerations:

$$W_{m+1} = W_m + \omega_m \varepsilon^\alpha \overline{X}'_m \qquad (4.2.2)$$

It is possible to remove all vectors X'_m, for which $W_{m+1} = W_m$ (that is $P(X'_m, \overline{W}_m) = \varepsilon^0$), from the learning sequence $S_{X'}$ for simplicity. $\tilde{X}_1, \tilde{X}_2, \ldots$ are the remaining members of the learning sequence, and $S_{\tilde{X}}$ is a reduced learning sequence. $S_{\tilde{W}}$ is a corresponding reduced sequence of the weighting vectors which is obtained according to the rule (4.2.2). Evidently, $\tilde{W}_p \ne \tilde{W}_q$, for $p \ne q$. The theorem will be proven, if we will prove that the sequence $S_{\tilde{W}}$ is finite. So from the assumption that $S_{\tilde{W}}$ is infinite, we have to get contradiction. The following equality is obtained from the condition $P(\tilde{X}_m, \tilde{W}_m) \ne \varepsilon^0$ and the rule (4.2.2):

$$\tilde{W}_{m+1} = \omega_{j_1} \overline{\tilde{X}}_1 + \omega_{j_2} \overline{\tilde{X}}_2 + \ldots + \omega_{j_m} \overline{\tilde{X}}_m, \qquad (4.2.3)$$

where $j_p = 1, 2, 3$ (corresponding to three cases for choice of ω); $p = 1, \ldots,$ m. Let without loss of generality $W_0 = 0$. According to the condition of the theorem such a vector $\tilde{W}$ exists that $P(X, \tilde{W}) = \varepsilon^0$ for all $X \in A'$. Let us fix this vector. Its scalar product with both parts of the equation (4.2.3) is equal to

$$(\tilde{W}_{m+1}, W) = (\omega_{j_1} \overline{\tilde{X}}_1, W) + \ldots + (\omega_{j_m} \overline{\tilde{X}}_m, W) \qquad (4.2.4)$$

or

$$(\tilde{W}_{m+1}, \overline{W}) = (\overline{\omega}_{j_1} \tilde{X}_1, \overline{W}) + \ldots + (\overline{\omega}_{j_m} \tilde{X}_m, \overline{W}). \qquad (4.2.5)$$

It follows from the condition $P(X,\widetilde{W}) = \varepsilon^0$ that

$$0 \leq \arg(\widetilde{X}_p, \overline{W}) < \frac{2\pi}{k}. \qquad (4.2.6)$$

Taking values of $\arg z$ from the interval $[-\pi, \pi]$ we obtain:

$$\arg(\overline{\omega}_1) = \arg(i\varepsilon^{-1}) = \frac{\pi}{2} - \frac{2\pi}{k} = -\frac{k-2}{2k}\pi - \frac{\pi}{k}, \qquad (4.2.7)$$

$$\arg(\overline{\omega}_3) = \arg(-i) = -\frac{\pi}{2} = -\frac{k-2}{2k}\pi - \frac{\pi}{k}, \qquad (4.2.8)$$

$$\arg(\overline{\omega}_2) = \arg(1) = 0 = \frac{\pi}{k} - \frac{\pi}{k}. \qquad (4.2.9)$$

Therefore:

$$-\frac{k-2}{2k}\pi = \arg(\overline{\omega}_3) + \frac{\pi}{k} \leq \frac{k-2}{2k}\pi, \qquad (4.2.10)$$

$$-\frac{k-2}{2k}\pi < \arg(\overline{\omega}_2) + \frac{\pi}{k} = \frac{\pi}{k} \leq \frac{k-2}{2k}\pi, \qquad (4.2.11)$$

if $k \geq 4$ (exactly such a case is more interesting);

$$-\frac{k-2}{2k}\pi < \arg(\overline{\omega}_1) + \frac{\pi}{k} = \frac{k-2}{2k}\pi. \qquad (4.2.12)$$

It is easy to obtain the following inequalities from the inequality (4.2.6) and equations (4.2.10)-(4.2.12):

$$-\frac{k-2}{2k}\pi \leq \arg\left[\omega_{j_p} e^{i\pi/k}(\widetilde{X}_p, \overline{W})\right] < \frac{k-2}{2k}\pi + \frac{2\pi}{k} = \frac{k+2}{2k}\pi, \qquad (4.2.13)$$

$$p = 1, 2, ..., m$$

It should be noted that the inequalities (4.2.13) are true not only for $\widetilde{X}_1, ..., \widetilde{X}_m$, but for all $X \in A'$. Let $\omega'_{j_p} = \overline{\omega}_{j_p} e^{i\pi/k}$, $p = 1, 2, ..., m$. Then the equality (4.2.5) and the inequalities (4.2.13) may be transformed respectively, to

$$(\overline{\widetilde{W}}_{m+1}, W)e^{i\pi/k} = \omega'_{j_1}(\widetilde{X}_1, \overline{W}) + ... + \omega'_{j_m}(\widetilde{X}_m, \overline{W}), \qquad 4.2.14)$$

$$-\frac{k-2}{2k}\pi \leq \arg\left[\omega^\odot_{j_p}(\widetilde{X}_p, W)\right] < \frac{k+2}{2k}\pi = \pi - \frac{k-2}{2k}\pi, \qquad (4.2.15)$$

$$p = 1, 2, ..., m$$

It is evident from the inequalities (4.2.15) that the complex numbers $\omega'_{j_p}(\widetilde{X}_p, \overline{W})$ belong   to   the   semi-plane

$$Q = \left\{ z \in C \mid -\frac{k-2}{2k}\pi \leq \arg(z) < \frac{k+2}{2k}\pi + \pi \right\}$$ of the complex plane, and

any of them does not belong to the ray

$$V = \left\{ z \in C \mid \arg(z) = -\frac{k-2}{2k}\pi + \pi \right\}.$$ Inequalities (4.2.15), as well as

(4.2.13) are not only true for $\tilde{X}_1, ..., \tilde{X}_m$, but also for all $X \in A'$ and with

any ω'_{j_p} (taking any of the values $\omega'_1 = \overline{\omega}_1 e^{i\pi/k} = i\varepsilon^{-1}e^{i\pi/k} = ie^{-i\pi/k}$,

$\omega'_3 = \overline{\omega}_3 e^{i\pi/k} = -ie^{i\pi/k}$, $\omega'_2 = e^{i\pi/k}$). Taking into account that the set A' is

finite and there is no point $\omega'_{j_p}(X,\overline{W}), X \in A'$ on the ray V, we can

conclude that a sector S exists with side V, angle value $\delta > 0$, belonging to

the semi-plane Q such, that it does not contain points like

$\omega'_{j_p}(X,\overline{W}), X \in A'$. δ may be found as follows:

$$\delta = \min_{X \in A'}\left\{ \frac{k+2}{2k}\pi - arg\left[\omega'_0(X,\overline{W})\right]\right\}, \text{ where } arg(\omega'_0) = \max_{j=1,2,3}(arg(\omega'_j)).$$

Let us denote $v = \exp\left[\left\{\frac{k-2}{2k}\pi + \frac{\delta}{2}\right\}i\right]$. The complex plane is rotated by

the mapping $z \to vz$ by such a way that the line L (which is bisector of the

sector S) across the real axis. The sector $Q\backslash S$ is mapped to the one belonging

to the upper semi-plane (over the real axis). Since all points

$v\omega'_{j_p}(X,\overline{W}), X \in A'$ belong to the sector $v[Q/S]$ (but not coincide with

$(0, 0)$) then $\min_{X \in A'; j=1,2,3} Im\left[v\omega'_j(X,\overline{W})\right] = a > 0$. It is easy to obtain the

following correspondence from the equation (4.2.14):

$$|(\tilde{W}_{m+1}, W)| =$$

$$= |(\tilde{W}_{m+1}, W)ve^{i\pi/\kappa}| \geq Im\left\{v\omega'^0_{j_1}(\tilde{X}_1, W) + ... + v\omega'^0_{j_m}(\tilde{X}_m, W)\right\} \geq ma.$$

Taking into account the last inequalities and Schwartz inequality (the

squared scalar product of the two vectors is less than or equal to the product

of the squared norms of these vectors) [Hoffman & Kunze (1971)], we

obtain $|\tilde{W}_{m+1}|^2 \geq \frac{a^2}{|W|^2}m^2$. Simultaneously the equalities

$\tilde{W}_{p+1} = \tilde{W}_p + \omega_{j_p}\tilde{X}_p$ (where ω_{j_p} is equal to 1, or ω_1, or ω_2) are also true

for all $p = 1, 2, ..., m$. The following equality is also true:

$$|\tilde{W}_{p+1}|^2 = |\tilde{W}_p|^2 + |\tilde{X}_p|^2 + 2\operatorname{Re}(\omega_{j_p}\tilde{X}_p, \overline{\tilde{W}_p}) =$$

$$= |\tilde{W}_p|^2 + |\tilde{X}_p|^2 + 2\operatorname{Re}(\overline{\omega}_{j_p}\tilde{X}_p, \overline{\tilde{W}_p}) \qquad (4.2.16)$$

According to our assumption $P(\tilde{X}_p, \overline{\tilde{W}_p}) \neq \varepsilon^0$. Therefore

$$\operatorname{Re}(\overline{\omega}_{j_p}\tilde{X}_p, \overline{\tilde{W}_p}) \leq 0. \qquad (4.2.17)$$

Really, let us consider all three cases for ω. In the case 1 $\arg(\varepsilon) \leq \arg\left[(\tilde{X}_p, \overline{\tilde{W}_p})\right] \leq \arg(\varepsilon^{k/4+1})$, and simultaneously the complex number $\overline{\omega}_{j_p}(\tilde{X}_p, \overline{\tilde{W}_p}) = \overline{\omega}_1(\tilde{X}_p, \overline{\tilde{W}_p}) = i\varepsilon^{-1}(\tilde{X}_p, \overline{\tilde{W}_p})$ has a non-positive real part. In the case 2 $\arg(\varepsilon^{k/4+2}) \leq \arg\left[(\tilde{X}_p, \overline{\tilde{W}_p})\right] \leq \arg(\varepsilon^{3k/4+1})$ and also $\operatorname{Re}\left[\overline{\omega}_{j_p}(\tilde{X}_p, \overline{\tilde{W}_p})\right] = \operatorname{Re}\left[(\tilde{X}_p, \overline{\tilde{W}_p})\right] \leq 0$. Finally, in the case 3

$$\arg(\varepsilon^{3k/4+1}) \leq \arg\left[(\tilde{X}_p, \overline{\tilde{W}_p})\right] \leq \arg(\varepsilon)$$

and $\overline{\omega}_{j_p}(\tilde{X}_p, \overline{\tilde{W}_p}) = \overline{\omega}_2(\tilde{X}_p, \overline{\tilde{W}_p}) = -i(\tilde{X}_p, \overline{\tilde{W}_p})$ also has a non-positive real part. Thus, the inequality (4.2.17) is true for all $p = 1, 2, \dots m$. Therefore from the equality (4.2.16) we obtain:

$$|\tilde{W}_{p+1}|^2 - |\tilde{W}_p|^2 \leq |\tilde{X}_p|^2, p = 1, 2, \dots \qquad (4.2.18)$$

In such a case, adding the left and the right side of the inequality (4.2.18) for $p = 1, 2, \dots, m$, we obtain the following:

$$|\tilde{W}_{m+1}|^2 \leq \sum_{p=1}^{m} |\tilde{X}_p|^2 = m(n+1). \qquad (4.2.19)$$

Evidently, the inequalities (4.2.15) and (4.2.19) are mutual contradictory for the great values of m (a, $|W|$ and $n+1$ are the constants according to the condition). So the sequence $S_{\tilde{W}}$ can't be infinite. That means convergence of the learning algorithm. Taking into account that there are exactly $k!$ permutations from the elements of the set $E_k = \{0, 1, \dots, k-1\}$, it is easy to conclude: the extreme upper estimate for the value of m_0 beginning, from which the sequence $S_{\tilde{W}}$ is expected to be stabilized, is the following: $m_0 \leq k!$. The theorem is proven.

Two important remarks should be given. 1) without loss of generality we can put $P[(0,0)] = 1$; 2). the theorem shows that there are different strategies of the learning corresponding to the different permutations R. The learning algorithm is independent of them and converges. The two strategies mentioned above are only the most general and natural.

3.   THE SECOND LEARNING RULE FOR THE MVN LEARNING ALGORITHM

The second learning rule for the MVN learning, which will be presented in this section, is the error-correction rule. The error-correction rule for the perceptron learning is described for example, in [Haykin (1994)]:

$$W_{m+1} = W_m + (q - s) X ,$$

where W_m and W_{m+1} are respectively, the current and the next weighting vectors, X is a current element (vector) of the learning set, q is a desired neuron's output, and s is an actual neuron's output. It is easy to convince that such a correction rule really moves the weighted sum to opposite domain.

Since the MVN implements a k-valued function, there are k equivalent domains (sectors in the complex plane), to which the weighted sum may fall. But only one of them corresponds to the desired neuron's output.

A generalization of the perceptron error-correction rule for the MVN learning has been proposed in [Aizenberg N. et al. (1995)]. Since MVN's output is a root of unity we obtain the following error-correction learning rule for MVN:

$$W_{m+1} = W_m + C_m (\varepsilon^q - \varepsilon^s) \overline{X} , \qquad (4.3.1)$$

where W_m and W_{m+1} are current and following weighting vectors, $\overline{X}$ is the vector of the neuron's input signals (complex-conjugated), ε is the primitive k^{th} root of unity (k is chosen from (2.2.2)), C_m is a scale coefficient, q is a number of the "correct" (desired) sector, s is a number of the sector, to which actual value of the weighted sum has fallen. The correction rule (4.3.1) has been proposed in [Aizenberg N. et al. (1995)], but with the constant normalized factor instead of C_m. Further experience showed that by changing of the scale factor the learning process could accelerate [Aizenberg N. et al. (1996 b)].

The role of scale coefficient C_m is the same that for the learning rule (4.2.1) (see the Section 4.2). It also should contain a multiplier $\dfrac{1}{(n+1)}$. It is evident from the following expression for the weighted sum corresponding to the weighting vector, which has been corrected according to the learning rule (4.3.1):

$$(X, \overline{W}_{m+1}) = (X, \overline{W}_m) + C_m (X, \overline{(\varepsilon^q - \varepsilon^s) \overline{X}}) =$$
$$(X, \overline{W}_m) + C_m (\varepsilon^q - \varepsilon^s)(X, X) = (X, \overline{W}_m) + (n+1) C_m (\varepsilon^q - \varepsilon^s).$$

We will return to the role of coefficient C_m in the Section 4.5, where the computational implementation of the learning process will be considered.

Theorem 4.3.1 (About the convergence of the learning algorithm with the learning rule (4.3.1)). If the learning subsets A_0, A_1, ..., A_{k-1} are k-separated then the learning algorithm with the learning rule (4.3.1) converges.

Proof. Let consider the new learning subsets $A'_j = (\varepsilon^{-j})A_j$, $j = 0, 1, ...,$ k-1. The set $A' = \bigcup_{0 \le j \le k-1} A'_j$ is a reduced learning set. We will say that this set is separable, if the sets A'_0, A'_1, ..., A'_{k-1} are k-separated. Thus, if A' is separable, it implies that a vector W exists such that $P(X', \overline{W}) = \varepsilon^0$ for all $X' \in A'$. The last condition is true if and only if the equation (4.1.1) is true ($P(X, \overline{W}) = \varepsilon^{\alpha_j}$). Therefore it is sufficient to prove the theorem for the reduced learning set. $S_{X'} = X'_1, X'_2,...$ is a learning sequence from the elements of the reduced learning set A', and $X'_m = \varepsilon^{-\alpha_j} X_m$, if $X_m \in A_j$, j = 0, 1, ..., k-1; $m = 0, 1, 2, ...$. Let for simplicity $C_m \equiv 1$. Let $P[(0, 0)]=1$. We can transform the rule (4.3.1) for the learning subsets A':

$$W_{m+1} = W_m + (1 - e^{(q-s)}) \overline{X}'_m,\qquad(4.3.2)$$

because the desired sector has number 0, and $\varepsilon^0 = 1$. We can remove the vectors X'_m, for which $W_{m+1} = W_m$, (that is $P(X'_m, \overline{W}_m) = \varepsilon^0$) from the learning sequence $S_{X'}$. Let $\tilde{X}_1, \tilde{X}_2,...$ are the rest members of the learning sequence, and $S_{\tilde{X}}$ is a reduced learning sequence. $S_{\tilde{W}}$ is a corresponding reduced sequence of the weighting vectors, which is obtained according to the rule (4.3.2). Evidently, $\tilde{W}_a \ne \tilde{W}_b$ for $a \ne b$. The theorem will be proven, if we will prove that the sequence $S_{\tilde{W}}$ is finite. Let $\tilde{W}_0 = ((0,0),....,(0,0))$ (without loss of generality). We obtain from the rule (4.3.2):

$$\tilde{W}_{m+1} = (1 - \varepsilon^{q_1-s_1}) \tilde{X}_1 + ... + (1 - \varepsilon^{q_m-s_m}) \tilde{X}_m.\qquad(4.3.3)$$

Since the learning sets A_0, A_1, ..., A_{k-1} are k-separated according to the condition, then a weighting vector W exists such that

$$P(X', \overline{W}) = \varepsilon^0 = 1\qquad(4.3.4)$$

for all $X' \in A'$. Let us evaluate the scalar product of this vector with both parts of equality (4.3.3) :

$$(\tilde{W}_{m+1}, W) = \left[((1 - \varepsilon^{q_1-s_1}) \tilde{\overline{X}}_1, W) + ... + ((1 - \varepsilon^{q_m-s_m}) \tilde{\overline{X}}_m, W) \right],$$

$$s_t, q_t = 0, 1, ..., k-1; \quad t = 1, 2, ..., m.\qquad(4.3.5)$$

Taking into account that the complex numbers $\varepsilon^{q_p - s_p}$ and $\varepsilon^{s_p - q_p}$ are mutual conjugate, and also taking into account the features of the scalar product within the unitary space, we can transform the last equality to the following:

$$(\overline{\widetilde{W}_{m+1}}, \overline{W}) = \left[((1 - \varepsilon^{s_1 - q_1}) \widetilde{X}_1, \overline{W}) + \ldots + ((1 - \varepsilon^{s_m - q_m}) \widetilde{X}_m, \overline{W}) \right]$$

(4.3.6a)

$s_t, q_t = 0, 1, \ldots, k - 1; \quad t = 1, 2, \ldots, m.$

If $s_p - q_p < 0$ for some p and q then it is always possible to reduce this difference *mod k*, so that it will be positive. So, let $r_p = s_p - q_p > 0$, $p = 1$, 2, ..., m. The last equality is transformed to:

$$(\overline{\widetilde{W}_{m+1}}, \overline{W}) = \left[((1 - \varepsilon^{r_1}) \widetilde{X}_1, \overline{W}) + \ldots + ((1 - \varepsilon^{r_m}) \widetilde{X}_m, \overline{W}) \right]$$

(4.3.6b)

$s_t, q_t = 0, 1, \ldots, k - 1; \quad t = 1, 2, \ldots, m.$

Evidently, from the equality (4.3.4) we obtain the following:

$$0 \le \arg(\widetilde{X}_p, \overline{W}) < 2\pi / \kappa, \, p = 1, 2, \ldots, m,$$

(4.3.7)

because W is a weighting vector for all $X' \in A'$ including all $\widetilde{X}$. On the other hand $|2\pi r_p / k - \pi| \le \pi$. If $r_p \ne 0$ then

$$\arg(1 - \varepsilon^{r_p}) = -2\pi r_p / k / 2 = -\pi r_p / k.$$

(4.3.8)

If $r_p = 0$ then $1 - \varepsilon^{r_p} = 1 - 1 = 0$. Taking into account that

$$\arg\left[(1 - \varepsilon^{r_p}) \widetilde{X}_p, \overline{W} \right] = \arg\left[(1 - \varepsilon^{r_p}) (\widetilde{X}_p, \overline{W}) \right],$$

from the inequality (4.3.7) and equality (4.3.8) we obtain the following inequality:

$$-\pi r_p / k \le \arg\left[(1 - \varepsilon^{r_p}) (\widetilde{X}_p, \overline{W}) \right] < -\pi r_p / k + 2\pi / k$$

or after simplification:

$$-\pi r_p / k \le \arg\left[(1 - \varepsilon^{r_p}) (\widetilde{X}_p, \overline{W}) \right] < (2\pi - \pi r_p) / k.$$

(4.3.9)

From the $r_p = (s_p - q_p) \bmod k$ it is evident that $\max(r_p) = k - 2$; $p = 1, 2, \ldots, m$. It is impossible that $\max(r_p) = k - 1$ because in such a case a current value of the weighted sum has already fallen into the desired sector, but we consider the reduced sequences $S_{\widetilde{X}}$ and $S_{\widetilde{W}}$, which correspond to the weighted sum getting into the "incorrect" sector only. Let us substitute r_p in the inequality (4.3.9) by its maximal value $k-2$:

$$-\pi(k - 2) / k \le \arg\left[(1 - \varepsilon^{r_p}) (\widetilde{X}_p, \overline{W}) \right] < (2\pi - \pi(k - 2)) / k,$$

or after a simplification:

$$\pi - 2\pi / k \le \arg\left[(1 - \varepsilon^{r_p})(\tilde{X}_p, \overline{W})\right] < \pi .$$

The last inequality shows that the complex numbers $(1 - \varepsilon^{r_p})(\tilde{X}_p, \overline{W})$ belong to the sector Q of the complex plane. This sector is within the upper and the left semi-planes (in relation to the real and imaginary axis), and between the rays corresponding to angles $\pi - 2\pi/k$ and π. The angle value of this sector is equal to $2\pi/k$. It should be noted that according to the condition $P(0, 0) = 1$ (see above) the complex number $(0, 0)$ does not belong to the sector Q. Evidently, $\operatorname{Im}(z) > 0, z \in Q, z \in C$. Let $a = \min\left\{\operatorname{Im}\left((1 - \varepsilon^{r_p})(\tilde{X}_p, \overline{W})\right)\right\}$. It is absolutely evidently that $a > 0$. We have to obtain an estimate for the absolute value of $(\tilde{W}_{m+1}, W)$. It is followed from the equalities (4.3.5) and (4.3.6 a-b) that

$$|(\tilde{W}_{m+1}, W)| = |(\tilde{W}_{m+1}, \overline{W})| . \tag{4.3.10}$$

Then we obtain the following:

$$|(\tilde{W}_{m+1}, \overline{W})| \ge \operatorname{Im}\left[((1 - \varepsilon^{r_1})\tilde{X}_1, \overline{W}) + \ldots + ((1 - \varepsilon^{r_m})\tilde{X}_m, \overline{W})\right] \ge ma, \tag{4.3.11}$$

and according to the Schwartz inequality (the squared scalar product of the two vectors is less than or equal to product of the squared norms of these vectors) [Hoffman & Kunze (1971)], the following inequality is true:

$$|(\tilde{W}_{m+1}, W)| \le \|\tilde{W}_{m+1}\| \, \|W\|, \tag{4.3.12a}$$

where $\|...\|$ is Euclidian norm, or in another form:

$$(\tilde{W}_{m+1}, W)^2 \le (\tilde{W}_{m+1}, \tilde{W}_{m+1})^2 (W, W)^2 . \tag{4.3.12b}$$

From the inequalities (4.3.11) - (4.3.12) we obtain the following estimate:

$$|\tilde{W}_{m+1}|^2 \ge (a^2 / |W|^2)m^2 . \tag{4.3.13}$$

On the other hand we can obtain another estimate. Indeed, let $1 - r_p = d_p, p = 1, 2, \ldots, m$. In such a case the equality (4.3.2) is transformed to: $\tilde{W}_{p+1} = \tilde{W}_p + d_p \tilde{X}_p, p = 1, 2, \ldots, m$. From the last equality we obtain the following:

$$\begin{aligned} |\tilde{W}_{p+1}|^2 &= |\tilde{W}_p|^2 + |X|_p^2 + 2\operatorname{Re}(d_p \tilde{X}_p, \tilde{W}_p) = \\ &= |\tilde{W}_p|^2 + |X|_p^2 + 2\operatorname{Re}(\overline{d}_p \tilde{X}_p, \tilde{W}_p). \end{aligned} \tag{4.3.14}$$

Taking into account that $P(\tilde{X}_p, \overline{W}_p) \ne \varepsilon^0 \ne 1$, we obtain

$$Re(\overline{d}_p \tilde{X}_p, \tilde{W}_p) \le 0, p = 1, 2, \ldots, m . \tag{4.3.15}$$

Then we obtain from the equality (4.3.14) and the inequality (4.3.15) the following:

$$|\widetilde{W}_{p+1}|^2 - |\widetilde{W}_p|^2 \le |\widetilde{X}|_p^2, \; p=1, 2, ..., m. \qquad (4.3.16)$$

Adding the left and the right sides of the inequality (4.3.16) for $p = 1, 2, ...,$ m, we obtain the following estimate:

$$|\widetilde{W}_{m+1}|^2 \le |\widetilde{X}_1|^2 + ... + |\widetilde{X}_m|^2 = m(n+1). \qquad (4.3.17)$$

The inequalities (4.3.13) and (4.3.17) are mutual contradictory for the great values of m (a, $|W|$ and $n+1$ are the constants according to the condition). Therefore the sequence $S_{\widetilde{W}}$ can't be infinite. That means a convergence of learning algorithm with the rule (4.3.1). Theorem is proven.

We have to conclude that different learning strategies can be considered. Other learning rules also may be considered. But two strategies considered in the Section 4.1 (respectively, a motion through learning subsets, and a motion through all learning set) are the most natural. Two linear learning rules (4.2.1) and (4.3.1) are very effective first of all from the point of view of their computing complexity. It should be noted that any of the considered learning rules have no significant advantages in comparison with another one. For example, the rule (4.3.1) does not contain an additional multiplication with ε^q, but at the same time learning with the rule (4.2.1) may converge faster for the non-smoothed functions. Therefore both learning rules have independent value and complement each other. We will consider some aspects of the computational implementation of the learning in the Section 4.5.

4. LEARNING ALGORITHMS FOR UNIVERSAL BINARY NEURON

We will consider in this Section only the UBN over the field of the complex numbers, as it was described in the Section 3.2. The learning algorithms for UBN over the finite algebras also may be considered, but at the same time the weights for implementation of Boolean functions on the UBN exactly over the finite algebras may be found very easily by sorting. On the other hand learning for UBN over the field of the complex numbers is important and interesting mathematical problem.

Thus, we will consider an UBN defined by the Definition 3.2.2 with the activation function (3.2.1). Two learning algorithms for UBN will be considered here. The first one is based on the Theorem 3.2.4, which shows that if some Boolean function $f(x_1,...,x_n)$ is P-realizable with the predicate P_B then some partially defined multiple-valued function $\widetilde{f}(x_1,...,x_n)$ is the m-valued threshold function. The weighting vector of such a partially defined m-valued threshold function will be simultaneously the weighting vector of the corresponding Boolean function. But a partially defined m-valued threshold function may be implemented using MVN. Since the learning algorithm for MVN has been already considered, and its convergence has been proven, it is possible to reduce the learning of UBN to learning of MVN. Another approach consists in direct learning of the UBN.

We will start from the first approach. Let us consider an example for the beginning.

Example 4.4.1. The activation function (the predicate P_B) is defined by the equality (3.2.1), $m=4$. As it was shown above (Example 3.1.1, Table 3.1.1) the XOR function of two variables is realisable on the single UBN with the weighting vector $(0, i, 1)$. From the Table 4.4.1 and Fig. 4.4.1 it is clear that simultaneously 4-valued partially defined (defined only on the Boolean combinations of variables) function $\tilde{f}(y_1, y_2)$ (or function $\tilde{f}(x_1, x_2)$) is 4-valued threshold function.

Table 4.4.1. Solution of the XOR problem on the single UBN with the weighting vector $(0, i, 1)$, and a correspondence with the partially defined four-valued function.

x_1	x_2	$z = w_0 + w_1 x_1 + w_2 x_2$	$P_B(z)$	XOR	$\tilde{f}(y_1, y_2)$	$\tilde{f}(x_1, x_2)$
1	1	$1 + i$	1	1	0	ε^0
1	-1	$-1 + i$	-1	-1	1	ε^1
-1	1	$1 - i$	-1	-1	3	ε^3
-1	-1	$-1 - i$	1	1	2	ε^2

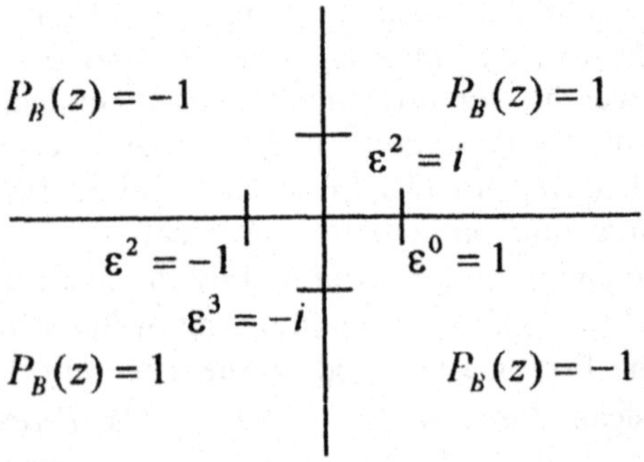

$$P_B(z) = -1 \qquad\qquad P_B(z) = 1$$
$$\varepsilon^2 = i$$
$$\varepsilon^2 = -1 \qquad\qquad \varepsilon^0 = 1$$
$$\varepsilon^3 = -i$$
$$P_B(z) = 1 \qquad\qquad P_B(z) = -1$$

Figure 4.4.1. Correspondence between predicates P and P_B ($m=4$) (see Example 4.4.1).

It will be very nice to find some universal approach for specific recoding of Boolean function to multiple-valued threshold function. Then the learning or synthesis algorithm should be applied and the problem is solved. Of course, it is attractive, but such a universal approach is still an open research problem. We would like to outline here some directions for a solution of the problem.

Let us consider the XOR problem. The XOR function (mod 2 addition) of n variables may be presented in the following way:

$$f(y_1, ..., y_n) = y_1 \oplus ... \oplus y_n , \qquad (4.4.1\ a)$$

$$f(x_1, ..., x_n) = x_1 \cdot ... \cdot x_n . \qquad (4.4.1\ b)$$

It should be remainded that y are the Boolean variables in the classical alphabet $\{0, 1\}$, and x are the Boolean variables in the alphabet $\{1, -1\}$. The correspondence between these alphabets is established by the equality (1.1.5). Let replace mod 2 addition in the equation (4.4.1 a) by decimal addition. The following multiple-valued function will be obtained as a result:

$$\tilde{f}(y_1, ..., y_n) = y_1 + ... + y_n , \qquad (4.4.2\ a)$$

$$\tilde{f}(x_1, ..., x_n) = e^{y_1 + ... + y_n} . \qquad (4.4.2\ b)$$

The function $\tilde{f}$ is a partially defined multiple-valued function. It is defined exactly on the Boolean combinations of variables: on the set $\tilde{E}_2^n \subset \tilde{E}_m^n$, where m is taken from the equality (3.2.1) (number of sectors, on which the complex plane has devised). The vector of values of the function $\tilde{f}$ is the following:

$$\tilde{f} =$$

$$= (0,1,1,2,1,2,2,3,2,3,3,4,3,4,4,5,...,...,...,n-2,n-1,n-1,n) \qquad (4.4.3)$$

It is clear from (4.4.3) that $\tilde{f}$ is a monotonic function for any number of variables. An application of the learning algorithm with the rule (4.2.1) or (4.3.1) to the partial-defined multiple-valued function (4.4.2 b) gives a quick convergence starting from any vector. A simple experiment shows that an estimate of the value of m in (3.2.1), which ensure convergence of the learning not more than after 10-100 iterations (few seconds on any slow computer) is $2n \le m \le 3n$. Moreover, it is easy to prove on the base for example, of the Theorems 2.2.1 - 2.2.4 that the function (4.4.2 b) of a form (4.4.3) is a partially defined multiple-valued threshold function. We tried to generalize an approach, by which the function (4.4.2) of a form (4.4.3) may be obtained from the function (4.4.1) [Aizenberg & Aizenberg (1994 b)]. Such a generalization is based on the following. The multiple-valued function (4.4.2 a) is a sum of Boolean variables. At the same time the mod 2 polynomial of XOR function contains exactly only Boolean variables (conjunctions of the zero rank). We assumed that it is always possible to obtain a multiple-valued threshold function from the mod 2 polynomial of Boolean function in the following simple way. It was shown in the Section 3.3 that mod 2 polynomial of a Boolean function may be obtained using the conjunctive (Reed-Muller) transformation over the field GF(2) (see (3.3.7)). Let f be a vector of values of a Boolean function (in the classical Boolean alphabet $\{0, 1\}$). Then

$$h = Kf \qquad\qquad (4.4.4\ a)$$

is a vector representing the coefficients of mod 2 polynomial of function f in Hadamard order (here K is a matrix of conjunctive transform (3.3.7), which is inverse to itself over the field GF(2)). Then the vector presenting the values of a monotonic multiple-valued function corresponding to Boolean function f may be obtained in the following way (here the matrix multiplication is considered over the field of the real numbers):

$$g = Kh, \qquad\qquad (4.4.4\ b)$$

where h is obtained according to (4.4.4 a).

The function g for the case, when f is the XOR function, coincides with the function (4.4.2). A parity of the values of function g obtained according (4.4.4 ab) always coincides with the parity of the values of function f (odd values of g correspond to "1s" of f, and even values of g correspond to "0s" of f). It is possible to apply the learning algorithm for MVN (or the algorithm of synthesis described in the Section 2.4) to the function g taking value of m in (3.2.1) from the estimate $m \geq 2n$. A convergence will be obtained for some multiple-valued functions corresponding to the non-threshold Boolean functions, but at the same time it will not be obtained for some other functions. Therefore an approach just described has a right for existence, but at the same time the learning of UBN, which is reduced to the learning of MVN still has the "white spots". It is disadvantage, but at the same time opens this field for further research.

Another approach to learning of UBN is its direct learning without recoding of a Boolean function implementing a corresponding input/output mapping. Such an approach may be based on the following considerations. An incorrect output of the UBN for some input vector X from the learning set means that a weighting sum has fallen into an "incorrect" sector. It is evident from the equality (3.2.1), which establishes the activation function of UBN, and the Fig. 3.2.1 that if the weighted sum gets into the "incorrect" sector, both of the neighborhood sectors are "correct" in such a case because $P_B(z_1) = P_B(z_2) = -P_B(z)$, $z \in (s)$ $z_1 \in (s+1)$, $z_2 \in (s-1)$ (Fig. 4.4.2).

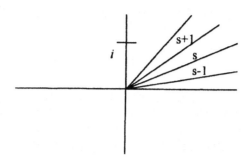

Figure 4.4.2 Definition of the "correct" sector for UBN learning

Thus, the weights should be corrected to direct the weighted sum into one of the neighborhood sectors. A natural choice of the "correct" sector (left or right) is based on the closeness of the current value of the weighted sum. To correct the weights we can use the same linear learning rules (4.2.1) and (4.3.1) that have been used for MVN learning.

An application of both MVN learning rules considered above means that a learning process for UBN is practically reduced to the learning process for MVN with truly binary inputs. Indeed, MVN is a neural element, which performs a correspondence between n neuron's inputs and single neuron's output described by multiple-valued threshold function, or function of m-valued logic (don't mix the k for MVN with m for UBN - compare the equalities (2.2.2) and (3.2.1)). If the learning process for some Boolean function has converged then the weighted sum corresponding to all the input combinations is located into the corresponding "correct" sectors on the complex plane. A problem of the convergence of learning according to the rules (4.2.1) and (4.3.1) is a problem of the stabilization of the sequence S_w of weighting vectors. Let S_x be a sequence of the input signals (input vectors). Let $A_0, A_1, ..., A_{m-1}$ are learning subsets-, so $A_j, j = 1, 2, ..., m-1$ contains the input vectors corresponding to the values of the weighted sums, which are finally (after finishing of the learning) located into the sector number j. If (2.2.2) and (3.2.1) are true for all the $A_0, A_1, ..., A_{m-1}$, we will say that the learning subsets $A_0, A_1, ..., A_{m-1}$ are m-separable.

The learning algorithm for UBN may be based on the following strategy. Let N be the cardinality of the learning set A, which consist of all the input vectors.

Step 1. The starting weighting vector W_0 is chosen as an arbitrary vector (e.g., its components may be the random numbers or may be equal to zero); $s=1; j=0$;

Step 2. Checking of the (3.2.2) for the element X_s of learning set:
> *if* (3.2.2) holds for it
> *then go to* the step 5
> *else begin* $j=1$; *go to* the step 3 *end*;

Step 3. A calculation of the angle distance between the arguments of the actual value of the weighted sum, the left border of right neighbor sector and the right border of the left one (see Fig. 4.4.2). The sector, to which the actual value of the weighted sum is closer to, should be chosen as "correct" one.
Go to the step 4.

Step 4. Obtaining of the vector W_{m+1} from the vector W_m by the learning rule (considered below);
> *go to* the step 2;

Step 5. $s = s+1$;
> *if* $s \leq N$
> *then go to* the step 2
> *else if* $j=0$
> > *then* the learning is finished successfully
> > *else begin* $s=1$; $j=0$; *go to* the step 2; *end*.

Here j is a flag of successful learning for some vector X_s. If current vector W satisfy (4.1.1) for all the vectors from learning set then $j=0$, if the vector W does not satisfy (4.1.1) for some vector X_s then $j=1$.

Both learning rules (4.2.1) and (4.3.1) considered above for MVN may be applied for implementation of the learning algorithm for UBN.

Let us consider the learning rule (4.2.1) applied to UBN. A correction of the weights is defined by the equality (4.2.1) without the complex conjugation of input vector [Aizenberg & Aizenberg (1994b)]:

$$W_{l+1} = W_l + \omega_l C_l \varepsilon^\alpha X, \qquad (4.4.5)$$

where X is the vector of the neuron's input signals (the current vector from the learning set), ε^α corresponds to the right bound of sector, to which we have to move the weighted sum, C_m is the scale coefficient, ω is the correction coefficient. Such a coefficient must be chosen from the point of view that the weighted sum should move exactly to the desired sector, or at least as close, as possible to it after the correction of the weights according to the learning rule (4.4.5). Evidently, for choice of ω only two cases (not three, as for MVN) should be considered because we always have to move only to the right or left neighbor sector. The detailed considerations for choice of ω

are given above in the Section 4.2. Evidently, the value of the coefficient ω for the UBN learning in (4.4.5) should be (i is imaginary unity):

$$\begin{cases} -i\varepsilon, & \text{if } Z \text{ must be "moved" to the right sector} \\ i, & \text{if } Z \text{ must be "moved" to the left sector} \end{cases} \qquad (4.4.6)$$

Theorem 4.4.1. (about a convergence of the learning algorithm with the learning rule (4.4.5)-(4.4.6)). If the Boolean function $f(x_1,...,x_n)$, which describes a mapping between inputs and output of the UBN, is implemented on the single UBN with activation function (3.2.1), then the weights $(w_0, w_1,..., w_n)$ that implement this function may be found by the learning algorithm based on the rule (4.4.5)-(4.4.6). In other words the learning algorithm based on the rule (4.4.5)-(4.4.6) converges after finite number of iterations.

Proof. This proof is similar to the proof of the Theorem 4.2.1 about convergence of the MVN learning with the similar learning rule. At the same time there are some differences, and we will give a separate complete proof of the Theorem 4.4.1 here. For a simplification, and without loss of generality we can prove the theorem only for the learning subsets

$$A_j' = \varepsilon^{-\alpha_j} A_j = \left\{ \varepsilon^{-\alpha_j} X_1^{(j)},...,\varepsilon^{-\alpha_j} X_{N_j}^{(j)} \right\}, \ j = 0,1, \ ..., \ m\text{-}1.$$ So we will

always suppose that the sector number 0 is a desired sector. The set $A' = \bigcup_{0 \le j \le m-1} A_j'$ is a reduced learning set. We will say that this set is separable,

if sets $A_0, A_1,..., A_{m-1}$ are m-separated. We will also mean for simplification that if weighted sum has fallen into the sector number j then $P(X, \overline{W}) = \varepsilon^j$ (compare with (4.1.1)). Thus, if A' is separable, it means that such a vector W exists that $P(X', \overline{W}) = \varepsilon^0$ for all $X' \in A'$. The last condition is true if, and only if (3.2.2) is true.

Let $S_{X'} = X_1', X_2',...$ be a learning sequence with the elements from the reduced learning set A'. It is evident that $X_l' = \varepsilon^{-\alpha_j} X_l$, if $X_l \in A_j$, $j = 0$, 1, ..., m-1; $l = 0, 1, 2, ...$. Let us suppose $C_l = 1$ in the equality (4.4.5) without loss of generality. We can transform the equality (4.4.6) taking into account the last considerations:

$$W_{l+1} = W_l + \omega_l \varepsilon^{\alpha} X_l' \qquad (4.4.7)$$

It is possible to remove all vectors X_l', for which $W_{l+1} = W_l$ (that is $P(X_l', \overline{W_l}) = \varepsilon^0$), from the learning sequence $S_{X'}$ for simplicity. $\tilde{X}_1, \tilde{X}_2,...$ are the remaining members of the learning sequence, and $S_{\tilde{X}}$ is the reduced learning sequence. $S_{\widetilde{W}}$ is the corresponding reduced sequence of the weighting vectors, which is obtained according to the rule (4.4.7). Evidently, $\widetilde{W}_p \neq \widetilde{W}_q$, for $p \neq q$. The theorem will be proven, if we will prove that the sequence $S_{\widetilde{W}}$ is finite. So from the assumption that $S_{\widetilde{W}}$ is infinite we have to get a contradiction. The following equality is obtained from the condition $P(\widetilde{X}_l, \widetilde{W}_l) \neq \varepsilon^0$ and the rule (4.4.7):

$$\widetilde{W}_{l+1} = \omega_{j_1} \overline{\widetilde{X}}_1 + \omega_{j_2} \overline{\widetilde{X}}_2 + \ldots + \omega_{j_l} \overline{\widetilde{X}}_l, \tag{4.4.8}$$

where $j_p = 1, 2$ (corresponding to the two cases for choice of ω); $p = 1, \ldots, l$. Let $W_0 = 0$. Such a vector W exists that $P(X, \widetilde{W}) = \varepsilon^0$ for all $X \in A'$ according to the condition of the theorem. Let us fix this vector, and evaluate a scalar product of both parts of the equality (4.4.8) with it:

$$(\widetilde{W}_{l+1}, W) = (\omega_{j_1} \overline{\widetilde{X}}_1, W) + \ldots + (\omega_{j_l} \overline{\widetilde{X}}_l, W) \tag{4.4.9}$$

or

$$(\overline{\widetilde{W}}_{l+1}, \overline{W}) = (\overline{\omega}_{j_1} \widetilde{X}_1, \overline{W}) + \ldots + (\overline{\omega}_{j_l} \widetilde{X}_l, \overline{W}) \tag{4.4.10}$$

The condition $P(X, \widetilde{W}) = \varepsilon^0$ involves

$$0 \leq \arg(\widetilde{X}_p, \overline{W}) < \frac{2\pi}{m}. \tag{4.4.11}$$

Taking the values arg z from the $[-\pi, \pi]$, we obtain:

$$\arg(\overline{\omega}_1) = \arg(i\varepsilon^{-1}) = \frac{\pi}{2} - \frac{2\pi}{m} = -\frac{m-2}{2m}\pi - \frac{\pi}{m}, \tag{4.4.12}$$

$$\arg(\overline{\omega}_2) = \arg(-i) = -\frac{\pi}{2} = -\frac{m-2}{2m}\pi - \frac{\pi}{m}. \tag{4.4.13}$$

Therefore

$$-\frac{m-2}{2m}\pi < \arg(\overline{\omega}_2) + \frac{\pi}{m} = \frac{\pi}{m} \leq \frac{m-2}{2m}\pi, \tag{4.4.14}$$

if $m \geq 4$ (exactly such a case is more interesting);

$$-\frac{m-2}{2m}\pi < \arg(\overline{\omega}_1) + \frac{\pi}{m} = \frac{m-2}{2m}\pi. \tag{4.4.15}$$

It is easy to obtain the following inequalities from the equality (4.4.11) and the equalities (4.4.12)-(4.4.13):

$$-\frac{m-2}{2m}\pi \leq \arg\left[\omega_{j_p} e^{i\pi/m}(\widetilde{X}_p, W)\right] < \frac{m-2}{2m}\pi + \frac{2\pi}{m} = \frac{m+2}{2m}\pi, \tag{4.4.16}$$

$$p = 1, 2, \ldots, l.$$

It should be noted that inequalities (4.4.16) are true not only for $\widetilde{X}_1, \ldots, \widetilde{X}_l$, but for all $X \in A'$. Let $\omega'_{j_p} = \overline{\omega}_{j_p} e^{i\pi/m}$, $p = 1, 2, \ldots, l$. Then the equality (4.4.10) and the inequalities (4.4.16) will be transformed to

$$(\overline{\overline{W}}_{l+1}, W)e^{i\pi/m} = \omega'_{j_1}(\widetilde{X}_1, \overline{W}) + \ldots + \omega'_{j_l}(\widetilde{X}_l, \overline{W}), \qquad (4.4.17)$$

$$-\frac{m-2}{2m}\pi \le \arg\left[\omega^0_{j_p}(\widetilde{X}_p, W)\right] < \frac{m+2}{2m}\pi = \pi - \frac{m-2}{2m}\pi, \qquad (4.4.18)$$

$p = 1, 2, \ldots, l$.

It is evident from the inequalities (4.4.18) that the complex numbers $\omega'_{j_p}(\widetilde{X}_p, \overline{W})$ belong to the semi-plane

$$Q = \left\{ z \in C \mid -\frac{m-2}{2m}\pi \le \arg(z) < \frac{m+2}{2m}\pi + \pi \right\}$$

of the complex plane, and any of them does not belong to the ray $V = \left\{ z \in C \mid \arg(z) = -\frac{m-2}{2m}\pi + \pi \right\}$. The inequalities (4.4.18), as well as (4.4.16), are true not only for $\widetilde{X}_1, \ldots, \widetilde{X}_l$, but for all $X \in A'$ and with any ω'_{j_p}, taking one of the values $\omega'_1 = \overline{\omega}_1 e^{i\pi/m} = i\varepsilon^{-1} e^{i\pi/m} = ie^{-i\pi/m}$, $\omega'_2 = \overline{\omega}_2 e^{i\pi/m} = -ie^{i\pi/m}$. Taking into account that the set A' is finite and there are no any point $\omega'_{j_p}(X, \overline{W})$, $X \in A'$ on the ray V, we can conclude that such a sector S exists with side V, angle value $\delta > 0$, belongs to the semi-plane Q, that it does not contain the points $\omega'_{j_p}(X, \overline{W})$, $X \in A'$. δ may be found like this: $\delta = \min_{X \in A'}\left\{ \frac{m+2}{2m}\pi - arg\left[\omega'_0(X, \overline{W})\right] \right\}$, where

$\arg(\omega'_0) = \max_{j=1,2}(\arg(\omega'_j))$. Let us denote $v = exp\left[\left\{ \frac{m-2}{2m}\pi + \frac{\delta}{2} \right\}i \right]$. The complex plane is rotated by the mapping $z \to vz$ in such a way that line L (which is a bisector of the sector S) across the real axis. Sector $Q\backslash S$ is mapped onto the sector belonging upper semi-plane (over the real axis). Since all the points $v\omega'_{j_p}(X, \overline{W})$, $X \in A'$ belong to the sector $v[Q/S]$ (but not equal to (0, 0)), then $\min_{X \in A'; j=1,2} Im\left[v\omega'_j(X, \overline{W}) \right] = a > 0$. It is possible to obtain now from the equation (4.4.17) the following:

$$|(\overline{\overline{W}}_{l+1}, W)| = |(\overline{\overline{W}}_{l+1}, \overline{W})ve^{i\pi/m}| \ge Im\left\{ v\omega'_{j_1}(\widetilde{X}_1, \overline{W}) + \ldots + v\omega'_{j_l}(\widetilde{X}_l, \overline{W}) \right\} \ge la.$$

Taking into account the last inequalities and Schwartz inequality [Hoffman & Kunze (1971)], we obtain the following: $|\widetilde{W}_{l+1}|^2 \geq \dfrac{a^2}{|W|^2} m^2$. The equalities $\widetilde{W}_{p+1} = \widetilde{W}_p + \omega_{j_p} \widetilde{X}_p$ (where ω_{j_p} is equal to ω_1 or ω_2) are true for all $p = 1, 2, ..., l$. The following equality is also true:

$$|\widetilde{W}_{p+1}|^2 = |\widetilde{W}_p|^2 + |\widetilde{X}_p|^2 + 2\operatorname{Re}(\omega_{j_p} \widetilde{X}_p, \widetilde{W}_p) =$$
$$= |\widetilde{W}_p|^2 + |\widetilde{X}_p|^2 + 2\operatorname{Re}(\overline{\omega}_{j_p} \widetilde{X}_p, \widetilde{W}_p) \qquad (4.4.19)$$

According to our assumption $P(\widetilde{X}_p, \overline{\widetilde{W}_p}) \neq \varepsilon^0$. Therefore

$$\operatorname{Re}(\overline{\omega}_{j_p} \widetilde{X}_p, \widetilde{W}_p) \leq 0. \qquad (4.4.20)$$

Indeed, let us consider both cases for ω. In the case 1 $\arg(\varepsilon) \leq \arg\!\left[(\widetilde{X}_p, \overline{W}_p)\right] \leq \arg(\varepsilon^{m/4+1})$, and simultaneously the complex number $\overline{\omega}_{j_p}(X_p, \overline{\widetilde{W}_p}) = \overline{\omega}_1(X_p, \overline{\widetilde{W}_p}) = i\varepsilon^{-1}(X_p, \overline{\widetilde{W}_p})$ has non-positive real part. In the case 2 $\arg(\varepsilon^{3m/4+1}) \leq \arg\!\left[(\widetilde{X}_p, \overline{W}_p)\right] \leq \arg(\varepsilon)$, and $\overline{\omega}_{j_p}(X_p, \overline{\widetilde{W}_p}) = \overline{\omega}_2(X_p, \overline{\widetilde{W}_p}) = -i(X_p, \overline{\widetilde{W}_p})$ also has non-positive real part. Therefore the inequality (4.4.20) is true for all $p = 1, 2, ... l$. Thus, from the equality (4.4.19) we obtain:

$$|\widetilde{W}_{p+1}|^2 - |\widetilde{W}_p|^2 \leq |\widetilde{X}_p|^2, \; p = 1, 2, ... \qquad (4.4.21)$$

Adding the left and the right side of the inequality (4.4.21) for $p = 1, 2, ..., l$, we obtain the following:

$$|\widetilde{W}_{l+1}|^2 \leq \sum_{p=1}^{m} |\widetilde{X}_p|^2 = l(n+1). \qquad (4.4.22)$$

Evidently, the inequalities (4.4.18) and (4.4.22) are mutual contradictory for the great values of l (a, $|W|$ and $n+1$ are the constants according to the condition). Therefore the sequence $S_{\widetilde{W}}$ can't be infinite. That means a convergence of the learning algorithm. Taking into account that there are exactly $m!$ permutations from the elements of the set $E_m = \{0, 1, ..., m-1\}$, it is easy to conclude: the extreme upper estimate for the value of l_0, beginning from which the sequence $S_{\widetilde{W}}$ is stabilized, is the following: $l_0 \leq m!$. The theorem is proven.

It is possible to prove a similar theorem taking instead of (4.4.6) the following rule for the choice of correction coefficient ω in (4.4.5) [Aizenberg & Aizenberg (1993, 1994 ab)]:

$$\begin{cases} 1, \text{ if } Z \text{ must be "moved" to the right sector} \\ \varepsilon, \text{ if } Z \text{ must be "moved" to the left sector} \end{cases}$$

The learning rule (4.3.1) should be also applied to the learning algorithm for UBN. It is taken the following form [Aizenberg I. (1997a)]:

$$W_{l+1} = W_l + C_l(\varepsilon^q - \varepsilon^s) \, X \, , \qquad (4.4.23)$$

where W_l and W_{l+1} are current and following weighting vectors, X is a vector of the neurons input signals, ε is aa primitive m^{th} root of unity (m is chosen from (3.2.1)), C_l is a scale coefficient, q is a number of the "correct" (desired) sector, s is a number of the sector, where the weighted sum is located. A number of the "correct" (desired) sector should be taken from the same considerations as for the learning rule (4.4.5):

$q = s - 1 \pmod{m}$, if Z is closer to $(s-1) \bmod m^{th}$ sector   (4.4.24 a)

$q = s + 1 \pmod{m}$, if Z is closer to $(s+1) \bmod m^{th}$ sector ,   (4.4.24 b)

where z is an actual value of weighted sum, s is a number of sector, where the weighted sum is located, q is a number of the "correct" (desired) sector.

Theorem 4.4.2. (about a convergence of the learning algorithm with the learning rule (4.4.23)-(4.4.24)). If the Boolean function $f(x_1,...,x_n)$, which describes a mapping between inputs and output of the UBN is implemented on the single UBN with activation function (3.2.1), then the weights $(w_0, w_1,..., w_n)$ that implement this function may be found by the learning algorithm based on the rule (4.4.23)-(4.4.24). In other words the learning algorithm based on the rule (4.4.23)-(4.4.24) is converges after finite number of iterations.

We will not present a proof because it is almost the same as for the Theorem 4.3.1 (see above) taking into account that here we deal with m-valued multiple-valued function. The interested reader can prove this theorem as exercise (taking into account also a notion of the m-separable learning sunsets).

A role of the scale coefficient C_m in the rules (4.4.5) and (4.4.23) is the same as in the rules (4.2.1), and (4.3.1) for the MVN learning. This coefficient may be changed during the learning process. C_m should contain

the multiplier $\dfrac{1}{(n+1)}$ because a weighting vector W is exactly an $(n+1)$-dimensional vector. We will return to the role of this coefficient in the Section 4.5.

5.   SOME COMPUTATIONAL ASPECTS OF THE LEARNING ALGORITHMS

The learning algorithms for MVN and UBN described above are adequate for solving of the learning problem. The error-correction learning rules (4.3.1) and (4.4.23) are simpler from the computing point of view, but at the same time the rules (4.2.1) and (4.4.5) give faster convergence for non-smooth functions.

Until now we presented many different aspects of MVN and UBN over the field of the complex numbers, but we did not speak about the computing complexity, especially related to the complex-valued arithmetic. As mentioned in the Chapter 1 that other authors also developed neurons with the complex arithmetic (e.g., [Leung & Haykin (1991)], [Georgiou & Koutsougeras (1992)], [Toth et. al. (1996)] should be noted). From our point of view the complex arithmetic is not a disadvantage, especially if it leads to the effective results in applications. Moreover, as it was shown above, only the complex arithmetic makes possible to develop a neuron with the universal functionality.

We do not discuss here the problems of VLSI implementation. It should be noted only that we do not see here some significant difficulties. For example, in [Aizenberg N. et. al. (1993)] it was proposed to implement the evaluation of the complex-valued weighted sum using two-layered DSP-architecture, and the evaluation of the activation function may be implemented using a simple table look-up.

But at the same time we would like to consider here some computing implementation aspects of learning in particular, and the MVN and UBN arithmetic in general.

Both learning rules (4.2.1) and (4.3.1) (for MVN, or (4.4.5) and (4.4.23) for UBN) may be implemented using direct application of the complex arithmetic with floating-point operations. A great advantage of motion around the unit circle makes a learning in our case very effective independently of starting values of the weights. A motion from any sector of the complex plane to any another one is the equivalent operation (one may compare with a motion corresponding to piece-wise linear models of [Moraga (1989)], [Si & Michel (1995)], which depends on the distance from the current value of weighted sum to the desired one. We omit here a comparison with the learning algorithms with nonlinear learning rules that e.g., lead to the solution of nonlinear programming problems). A significant acceleration of learning and all the computations related to MVN and UBN havs been proposed in [Aizenberg N. et al. (1995 a b)], [Aizenberg N. et. al. (1996 a)].

We would like to show here, how it is possible to get 15-17-times acceleration of the computations, first of all in the learning process. By avoiding computations with the floating-point numbers and operations. All the weights and roots of unity may be presented by the integer 32-bit numbers respectively, per real and imaginary part. The weighted sum may be presented by the integer 64-bit numbers respectively, per real and imaginary part. For transformation of the real values into integer ones we will use a scale factor D, which is equal to some power of 2 (from 8 till 16384). In the other words, if a is a real number, which presents real or imaginary part of the complex number, we will use integer

$$b = \text{Round}(Da) \qquad (4.5.1)$$

instead of a in all the computations. This modification gives high acceleration of the computations that are defined by the equalities (4.2.1), (4.3.1), (4.4.5), (4.4.23) and (2.2.1), (3.2.2). It is also a good mean to break errors connected with the computations with the irrational numbers. After convergence of the learning it is possible to reduce the weights to the 16-bit numbers for the real and imaginary part because higher precision is needed only during the learning process. Often it is possible to reduce a precision for the weights even to 12 bits for the real and imaginary part. The higher precision of the weights is very important exactly during the learning process, especially in the case, when the weighting sum for some combinations of variables is very close to the border between sectors. Without this precision mistakes in the computation of the neuron's output are possible. Of course, rejection of the floating-point operations gives a significant acceleration of all the computations, especially related to the learning process.

Another effective degree of freedom for control the learning is concentrated in the scale factor C_m (see the learning rules (4.2.1), (4.3.1), (4.4.5), (4.4.23)). As it was mentioned above C_m should contain the multiplier $\dfrac{1}{(n+1)}$, where n is a number of the neuron inputs. Therefore it may be presented as

$$C_m = \frac{1}{(n+1)} \tilde{C}_m . \qquad (4.5.2)$$

A starting value of the multiplier $\tilde{C}_m$ may be equal to 1, but changing the value of $\tilde{C}_m$ (its increasing, or decreasing, e.g., with a 2-times step) during the learning process it is often possible to accelerate a convergence of the learning algorithm. Such a change of $\tilde{C}_m$ is especially effective when the weighted sum is close to the border of sectors on some number of input combinations.

Another modification aims at a faster evaluation of the neuron output, when k (the value of multiple-valued logic in (2.2.2)), or m (the number of sectors, into which the complex plane has been devided by the activation function of UBN (3.2.1)) is a multiple of 8. It is possible to get 8-times reduction of the computing complexity of MVN's activation function P defined by (2.2.2) (and computing complexity of UBN's activation function P_B defined by (3.2.1)) exactly for such values of k and m. This is very important, because for example, for solution of image processing problems we will deal with k=256. An approach that is proposed here makes it possible to evaluate y (neuron's output in the integer form corresponds to the complex ε^y) very easily. Let $z = (\text{Re}(z), \text{Im}(z))$ be the complex number, which represents the weighted sum and

$$\varphi = \begin{cases} abs(\text{Re}(z))/abs(\text{Im}(z)) & \textit{if } abs(\text{Re}(z)) > abs(\text{Im}(z)) \\ abs(\text{Im}(z))/abs(\text{Re}(z)) & \textit{if } abs(\text{Im}(z)) < abs(\text{Re}(z)) \end{cases}. \qquad (4.5.3)$$

We will directly evaluate value of y using a table, which always may be created in advance (instead of evaluation of arg z and comparisons of angles to obtain the neuron's output y):

$$T(l) = tg(2\pi l/k), \text{ where } l=0,1,...,(k/8)\text{-}1. \qquad (4.5.4)$$

It is evident from the (4.5.3), (4.5.4), and Fig. 4.5.1 that such a value of l may be found that $T(l) \le \varphi \le T(l+1)$. Taking into account the sign and the value of the ratio $\text{Im}(z)/\text{Re}(z)$ it is easy to find a number s of the subsector, where z is located. Evidently, that $y=s(k/8)+l$; $s=0, 1, ..., (k/8)\text{-}1$ (k in the last equation and (4.5.4) may be replaced by m from (3.2.1), and the same considerations hold).

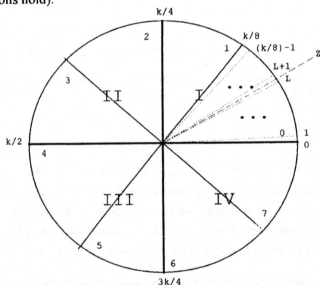

Figure 4.5.1 Evaluation of the activation function for the number of sectors devised on 8.

The computing implementation of the learning algorithm, which has been just considered, makes possible to work with such large values of k and m as e.g., 16384 without any computing problem. On the other hand computations with large values of k and m with floating-point numbers technique involves reduction of the computing speed and problems of the accumulation of errors connected with the irrational numbers expressed via $\pi, \exp$, etc.

Let us consider an example of the implementation of the learning algorithm for UBN with the learning rule (4.4.23) using the technique just described. The following weighting vectors were obtained for the XOR function of different number n of variables. The learning process started from the random vector (m is a number of sectors on the complex plane in (3.2.1)). Number of iterations needed for convergence of the learning is equal respectively, 5 ($n=3$) , 35 ($n=8$), 50 ($n=9$).

$n = 3, m=6,$ $W=(\,(0.0,\ 0.0), (\,-20.47, 0.0), (0.05, -35.5), (20.5, 0.0)\,)$
$n = 8, m=16,$ $W=(\,(99.7, -165.9), (3.4, 57.3), (-0.10, 66.2), (-1.4, -58.8),$
$(-92.4, -1.3), (-1.4, -60.1), (-94.4, -3.4), (90.2, 2.5), (1.2, 57.2)\,)$
$n = 9, m=22,$ $W=(\,(8.2, 67.9), (-0.5, -16.7), (0.06, -16.8), (-0.05, -16.7),$
$(-0.4, -17.6), (-0.3, -17.1), (-0.7, -16.6), (21.9, 1.3), (70.1, 1.6),$
$(-0.4, -16.6)\,)$.

6.   CONCLUSIONS

The learning algorithms for MVN and UBN have been considered in this Chapter. Two linear rules for MVN learning were proposed. Respectively, two fundamental theorems about convergence of the learning algorithm for both of learning rules have been proven. Two different strategies for MVN learning were considered. It was shown that the UBN learning might be reduced to the MVN learning. Two linear learning rules for UBN learning have been proposed, and corresponding theorems about convergence of the learning algorithm have been proven. Some computing aspects of the learning were considered. It was shown that avoiding computations with the floating-point numbers might get significant acceleration of the learning process.

Chapter 5

Cellular Neural Networks with UBN and MVN

Cellular Neural Networks (CNN) with UBN and MVN as the basic neurons (respectively, CNN-UBN and CNN-MVN) are considered in this chapter. A brief introduction to the CNN, and an observation of the related applied problems are presented. It is shown that the use of MVN and UBN as the basic CNN neurons may extend their functionality by implementing mappings described with non-threshold Boolean and multiple-valued threshold functions. Some problems of image processing, which may be effectively solved using CNN-UBN and CNN-MVN are considered: precise edge detection, edge detection by narrow direction, impulsive noise filtering, multi-valued nonlinear filtering for noise reduction and frequency correction (extraction of image details).

1. TRADITIONAL CNN, CNN-UBN AND CNN-MVN

CNN have been introduced in [Chua & Yang (1988)]. This paper has initiated the development of a new field in neural networks and their applications. CNN have become a very effective method for solving many different image processing and recognition problems. CNN development have moved away during last nine years from the simple applications of binary image processing (a lot of papers in [CNNA (1990)]) towards the implementation of different nonlinear filters (e.g., [Rekeczky et. al. (1996)] and color image processing [Lee & Pineda (1996)]. The appearance of CNN moved forward a work in three fields: non-linear dynamic systems, neural networks and image processing. We omit here the first field, and consider the second and the third ones.

A revolutionary role of the CNN in neural networks is first of all in a local connectivity. The neural networks considered before the paper [Chua &

Yang (1988)] usually where the fully- connected networks (e.g., Hopfield network), or multi-layered neural networks with full connections between neurons of neighbor layers (e.g., multi-layered perceptron). A CNN conception supposes a cellular structure of the network: each neuron is connected only with the neurons from its nearest neighborhood (see Fig. 5.1.1). It means that the corresponding inputs of each neuron are connected with outputs of neurons from the nearest *rxr* neighborhood, and on the other hand, the output of each neuron is connected with the inputs of neurons from the same neighborhood. The neurons of such a network are also often called cells.

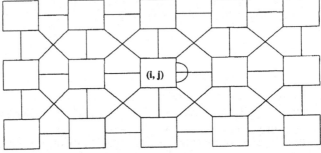

Figure 5.1.1. CNN of a dimension 3x5 with the local connections in a 3x3 neighborhood: each neuron is connected with 8 neurons around it and with itself

There is solid progress in the VLSI implementation of the proposed network. On the other hand a huge field of the applications of CNN in image processing has been opened. It was natural because CNN is a two-dimensional structure, and image processing is dealing with two-dimensional objects. Moreover, since many image processing algorithms may be reduced to the linear convolution with window in spatial domain, they could be implemented using CNN. A high productivity of the idea proposed in [Chua, Yang (1988)] is confirmed by hundreds of papers presented beginning from 1990 at five IEEE International workshops on CNN and their applications [CNNA (1990), (1992), (1994), (1996), (1998)]. An intensive developing of CNN theory (e.g., [Roska & Vandewalle (1993)]) initiated an idea of CNN universal machine (Roska & Chua (1993)] and their applications (many papers on this subject may be found e.g., in [CNNA (1996), (1998)]. Additionally to the references mentioned above the following CNN applications to image processing should be noted among other: implementation of morphological operations [Zarandy et al. (1996)], image segmentation [Sziranyi et. al. (1996)], [Vilarino et. al. (1998)], image halftoning [Crounse et. al. (1998)].

Usually two types of CNN are distinguished: *Continuous-time CNN* (or *classical CNN - CTCNN)* and *Discrete-time CNN (DTCNN)*.

CTNN were introduced in [Chua & Yang (1988)]. The dynamics of CTNN cell is described by the following equation:

$v(i, j) =$

$$= F\left[-z(i, j) + \sum_{k=-r}^{r}\sum_{l=-r}^{r}(A(k,l)v(i+k, j+l)) + \sum_{k=-r}^{r}\sum_{l=-r}^{r}(B(k,l)u(i+k,j+l) + I\right]^{(5.1.1)}$$

The dynamics of DTCNN cell introduced in [Harrer & Nossek (1992)] is described by the next equation:

$v(i, j, t) =$

$$= F_d\left[\sum_{k=-r}^{r}\sum_{l=-r}^{r}(A(k,l)v(i+k, j+l, t-1)) + \sum_{k=-r}^{r}\sum_{l=-r}^{r}(B(k,l)u(i+k, j+l, t-1)) + I\right].\quad (5.1.2)$$

Here (i, j) is the number of ij^{th} neuron, v is the neuron's output, u is the neuron's input, z is the neuron's state, r is the size of a neuron's nearest neighborhood $(r \times r)$, t, $t+1$ are the time slots t and $t+1$, respectively, A, B are the $r \times r$ matrixes, which define synaptic weights (feedback and control templates respectively), I is a bias, respectively F and F_d are the activation functions of a CTCNN and DTCNN neuron. For the CTCNN an activation function is the following:

$$F(z) = \frac{|z+1|-|z-1|}{2}. \quad (5.1.3)$$

For the DTCNN an activation function is the following:

$$F_d = \text{sgn}(z) \quad (5.1.4)$$

Actually, the DTCNN (but without such a term) were also considered in [Hao et. al. (1990)].

The following remark should me made. A nonlinear D-template for the implementation of nonlinear filters (e.g., median and rank-order ones) has been introduced in [Shi (1994), Rekeczky et. al. (1996]). The equation (5.1.1) is transformed in such a case as follows:

$$v(i, j) = F\left[-z(i, j) + \sum_{k=-r}^{r}\sum_{l=-r}^{r}(A(k,l)v(i+k, j+l)) + \right.$$

$$\left. + \sum_{k=-r}^{r}\sum_{l=-r}^{r}(B(k,l)u(i+k,j+l)) + \sum_{k=-r}^{r}\sum_{l=-r}^{r}\left(\hat{D}(k,l)\cdot\Delta V_{uzv}(t)\right) + I\right]$$

where $\hat{D}(k,l)$ is the generalized nonlinear term applied to ΔV, the voltage difference of either the input, state or output values.

The activation function (5.1.1) is a piece-wise linear function (see Fig. 1.1.2b). The activation function (5.1.2) is the simple *sign* function (see Fig. 1.1.2a) , therefore DTCNN is based on the threshold neurons. Evidently, the linear part of the function (5.1.3) ensures implementation of any algorithm of linear 2-D filtering in spatial domain, which is reduced to the linear convolution with the weighting window, on the CTCNN cell. DTCNN ensures implementation of binary image processing algorithms, which may be described by threshold Boolean functions.

Of course, the CTCNN and DTCNN open many possibilities to solve the different problems of image processing. Indeed, the formulas (5.1.1)-(5.1.3) enable an implementation of linear and nonlinear filters in spatial domain. Moreover it is a good way to implement the image processing algorithms described by threshold Boolean functions. On the other hand two problems should be mentioned here. The first one is that a derivation of corresponding templates is necessary for the implementation of any algorithm. Such a derivation may be achieved by several ways [Nossek (1996)]: learning (finding by some learning algorithm of the weighting templates performing the desired global mapping from input state to the corresponding desired output); design (e.g., programming the network to have some desired fixed points); mapping (e.g., simulation of biological phenomenon based on a set of equations). A problem is that many learning and design algorithms for CNN (see e.g., [Nossek (1996)] for observation) often are reduced to the solution of differential equations, which may be a specific computing problem. The second problem is that the DTCNN can't be used anywhere for an implementation of the processing algorithms described by non-threshold Boolean functions. The non-threshold function can't be implemented with any learning algorithm on the threshold neural element. But exactly the threshold neuron is a basic element of DTCNN. These two problems should not be considered as a disadvantage of CNN. Indeed, many interesting applied problems were solved within CTCNN and DTCNN traditional conception.

But the following ways for development of CNN paradigm are natural among other: 1) increasing of CNN functionality by using of MVN and UBN as basic neurons in the cellular network with the architecture presented in Fig. 5.1.1. This brings us to the CNN based on MVN and UBN (CNN-MVN and CNN-UBN); 2) A search for the new problems, which could not be solved within traditional CNN conception, but may be solved on CNN-MVN and CNN-UBN.

CNN-MVN have been proposed in [Aizenberg & Aizenberg (1992), (1993)] as a generalization of the associative memory proposed in [Hao & Vandewalle (1990)] for binary images on the case of gray-scale images. CNN-UBN have been proposed in [Aizenberg & Aizenberg (1993), (1994)] as instrument for solving the edge detection problem. Then a concept of

CNN-MVN and CNN-UBN has developed. The following results have been obtained and will be discussed below: impulsive noise filtering [Aizenberg N. et al. (1996)], an application of the CNN-UBN to small-detailed image processing [Aizenberg I. (1997a)], [Aizenberg & Aizenberg (1997)], multi-valued nonlinear filtering [Aizenberg I. (1997 b, c)], [Aizenberg I. & Vandewalle (1997)], general solution of precise edge detection problem [Aizenberg I. et. al. (1998a)], solution of the super resolution problem [Aizenberg I. & Vandewalle (1998)], [Aizenberg I. et. al. (1998)].

An efficiency of CNN-MVN and CNN-UBN is based on the following observations, made in the Chapters 1-4: 1) huge functionality (because of universal functionality of UBN, and practically universal functionality of MVN); 2) quickly converged learning algorithms; 3) use of the non-linearity of MVN activation function (2.2.2) for developing of nonlinear filters.

Let now us define CNN-MVN and CNN-UBN. Consider a CNN with the obvious local connection structure (see Fig. 5.1.1) of a dimension N x M. Consider MVN or UBN as a basic neuron of such a network. Evidently, each cell of CNN-MVN (or CNN-UBN) will perform the following correspondence between neuron's inputs and output:

$$Y_{ij}(t+1) = P\left[w_0 + \sum_{k=-r}^{r}\sum_{l=-r}^{r} w_{kl}^{ij} x_{kl}^{ij}(t) \right], \qquad (5.1.5)$$

where: ij are the two-dimensional coordinates (number) of a neuron; Y_{ij} is the neuron's output; w_{kl}^{ij} is the synaptic weight corresponding to the input of ij^{th} neuron, to which the signal from the output of kl^{th} neuron is transmitted (one may compare with the elements of B-template in the traditional CNN (see (5.1.1), and (5.1.2)); x_{kl}^{ij} is the input of ij^{th} neuron, to which the signal from the output of kl^{th} neuron is transmitted; P is the activation function of the neuron, which is defined by the equation (2.2.2) for MVN, and by the equation (3.2.1) for UBN; $\sum_{m=-r}^{r}$ means that each neuron is only connected with the neurons from its nearest r-neighborhood, as usual for CNN. Single weighting template of CNN-UBN and CNN-MVN may be written in the following form:

$$w_0; \quad W = \begin{pmatrix} w_{i-r,j-r}^{ij} & \cdots & w_{i-r,j+r}^{ij} \\ \cdots & w_{ij}^{ij} & \cdots \\ w_{i+r,j-r}^{ij} & \cdots & w_{i+r,j+r}^{ij} \end{pmatrix},$$

where W is rxr matrix.

What is common and what is different with the traditional CNN? First of all the cellular architecture of network is an important common point. It is evident, and it will be shown below on many examples, that CNN-MVN and CNN-UBN are oriented forwards solving the same problems of image processing and recognition, as traditional CNN. The main difference between the traditional CNN, and CNN-MVN, CNN-UBN is the use of complex-valued weights in the last ones. At first sight it is disadvantage. But at the same time CNN-UBN and CNN-MVN use a single complex-valued weighting template, and the traditional CNN uses two real-valued weighting templates. So, from the point of view of the weights complexity the traditional CNN, and CNN-UBN, CNN-MVN are equivalent. Since this book is devoted to theoretical aspects of MVN, UBN, and their applications, we will not consider here the VLSI implementation. This will be a nice subject for the future work. But it is not a fundamental problem from our point of view. It is absolutely evident that complex-valued arithmetic may be implemented on two-layered traditional CNN (such an approach has been proposed in [Toth et al. (1996)]). A value of activation function corresponding to the current value of the weighted sum may be obtained with the table that should be created in advance (see the Section 4.5). This idea has been developed in [Aizenberg N. et. al. (1993), (1995)].

Thus, to conclude this observation of CNN, it should be mentioned that CNN-UBN and CNN-MVN are not alternative to the traditional CNN. They may be considered as a nice supplement, which gives high efficiency in applications (we will consider them in the following Sections of this chapter). On the other hand an elaboration of new image processing algorithms has been initiated by the conception of CNN-UBN and CNN-MVN.

2.   PROBLEM OF PRECISE EDGE DETECTION, AND FILTERING BY BOOLEAN FUNCTIONS: SOLUTION USING CNN-UBN

Edge detection is a very important image processing problem. There are many different edge detection algorithms. Some of them are classical (Laplacian, Sobel, Prewitt - see [Pratt (1978)]) and well known. Several templates for traditional CNN have been developed (e.g., [Chua & Yang (1988)], [Nossek (1996)]). All of them have a common and significant disadvantage: they are usually sensitive to great brightness jumps only, and can't ensure detection of the edges corresponding to the small brightness jumps and details of a complicated form. Some methods of preliminary amplification of the brightness jumps are considered e.g., in [Argyle (1971)], [Pratt (1978)]. But such an approach could not be recognized as objective because it strongly depends on parameters, which are set by user. Different parameters lead to different results. The different methods of thresholding in combination with linear and nonlinear algorithms using in edge detection have the same disadvantage (the pioneer papers in this field are [Rosenfeld (1970)], [Rosenfeld & Thurston (1971)]). Different values of threshold, that may be chosen, lead to different results. We will not analyze here in detail the classical approaches. The reader who is interested is refered e.g., to [Pratt (1978)], [Sun & Venetsanopoulos (1988)]. We would like to add only that the edge detection on the gray-scale images using the traditional CNN beginning from [Chua & Yang (1988)] also is usually reduced to the thresholding of the image, and then to the processing of the obtained binary image. The different results are obtained again with the different threshold values. Moreover, non-contrast edges of some important details may be lost when the chosen value of the threshold is unlucky.

A natural question, is it possible to carry out an objective edge detection algorithm, which will detect all edges independently of image contrast? Is it possible for gray-scale images? For binary images? Our answer is positive. An approach, which will be presented here, has been proposed in [Aizenberg & Aizenberg (1993)], and then developed in [Aizenberg & Aizenberg (1994)], [Aizenberg I. (1997a)], [Aizenberg I. et. al. (1998a)].

Let consider the edge detection on binary images. To understand our approach, first of all it is necessary to have a common understanding on the notion of an edge? The following key items should be taken into account. 1) When we say that the edge in some pixel of binary image is detected, it means that the brightness value in such a pixel differs from brightness value at least of one of other pixels from the 3 x 3 window around it[3]. 2) The upward jumps of brightness values (from 0 to 1, if 1 is the value of brightness in the analyzed pixel), and the downward ones (from 1 to 0, if 0 is the brightness value in the analyzed pixel) are not equivalent in general. 3) It follows from the item 2 that the following three variants of edge detection should be considered to obtain the most general picture of the jumps of brightness:

1. Separate detection of edges corresponding to upward jumps;
2. Separate detection of edges corresponding to downward jumps;
3. Joint detection of edges corresponding to upward and downward jumps;

It should be mentioned (and it is very important) that the classical edge detection algorithms (Sobel, Laplace, Prewitt, etc.) do not differentiate between upward and downward jumps of brightness. It means that actually they are handling only the variant 3. The reader will convince below that the

---

[3] If the isolated pixels should be removed from the consideration then the following supplement to this condition should be added: the brightness value in such a pixel is equal to the brightness value at least of one of other pixels from the 3 x 3 window around it. We will not consider this case further.

variant 3 (also as the mentioned classical algorithms) leads to the "double" detection of edges in many pixels. As a result the corresponding edges will be "bold", and not precise. Such an effect should be classified as a great disadvantage, especially for gray-scale images (we will move from the binary to gray scale images little bit below).

Evidently, any of three mentioned variants of edge detection may be described by the corresponding Boolean function of nine variables:

$$f\begin{bmatrix} x_1 & x_2 & x_3 \\ x_4 & x_5 & x_6 \\ x_7 & x_8 & x_9 \end{bmatrix} = f(x_1, x_2, x_3, x_4, x_5, x_6, x_7, x_8, x_9).$$

Indeed, we consider a 3 x 3 window around the analyzed pixel including itself. It means that we have to analyze exactly 9 input values. Our Boolean functions should be equal to 1, if the edge is detected, and to 0 otherwise (respectively, to -1, and to 1, if the Boolean alphabet {1, -1} is used). It is also evident that CNN is an ideal structure for the implementation of the corresponding Boolean functions, and therefore, for the solution of edge detection problem. Of course, the corresponding Boolean function must be implemented on the CNN basic neuron (the reader will convince that a lot of functions that describe considered problems are not threshold and therefore could not be implemented on CTCNN, or DTCNN, but may be easily implemented on CNN-UBN).

It is also possible to consider the problem of edge detection by narrow direction at the same description: representation by Boolean function, and learning for an implementation of the corresponding function using CNN. A solution of this problem will practically complete the solution of the precise edge detection problem, and its CNN-implementation. It will be a very important instrument for processing first of all the small-detailed SAR-images (especially, for a detection of the sea streams through the edge detection "by direction").

Let consider the Boolean functions corresponding to the variants 1-3 distinguished above. Thus, we will analyze a 3 x 3 window around each pixel.

Variant 1. *Separate detection of edges corresponding to upward brightness jumps.*

The edge should be detected if, and only if the brightness in the central pixel of a 3 x 3 window is equal to 1, and brightness of the at least one other pixel within the same window is equal to 0. This process is described by the following Boolean function

$$f\begin{bmatrix} x_1 & x_2 & x_3 \\ x_4 & x_5 & x_6 \\ x_7 & x_8 & x_9 \end{bmatrix} = x_5 \& (\bar{x}_1 \vee \bar{x}_2 \vee \bar{x}_3 \vee \bar{x}_4 \vee \bar{x}_6 \vee \bar{x}_7 \vee \bar{x}_8 \vee \bar{x}_9). \qquad (5.2.1)$$

According to (5.2.1) edge is detected, if $x_5 = 1$, and at least one of other variables is equal to zero. It means that the function (5.2.1) detects the edges corresponding to the upward brightness jumps. Let consider several examples.

The edge will be detected by the function (5.2.1) e.g., for the following input values:

$$f\begin{bmatrix} 0 & 0 & 0 \\ 0 & 1 & 0 \\ 0 & 0 & 0 \end{bmatrix} = 1; \quad f\begin{bmatrix} 0 & 0 & 0 \\ 0 & 1 & 0 \\ 0 & 1 & 1 \end{bmatrix} = 1; \quad f\begin{bmatrix} 0 & 0 & 0 \\ 0 & 1 & 0 \\ 1 & 0 & 0 \end{bmatrix} = 1 \quad .$$

The edge will not be detected by the function (5.2.1) e.g., for the following input values:

$$f\begin{bmatrix} 1 & 1 & 1 \\ 1 & 1 & 1 \\ 1 & 1 & 1 \end{bmatrix} = 0; \quad f\begin{bmatrix} 0 & 0 & 0 \\ 0 & 0 & 0 \\ 1 & 0 & 0 \end{bmatrix} = 0; \quad f\begin{bmatrix} 1 & 0 & 1 \\ 1 & 0 & 1 \\ 0 & 1 & 1 \end{bmatrix} = 0.$$

Variant 2. *Separate detection of edges corresponding to downward brightness jumps.*

The edge should be detected if, and only if, the brightness in the central pixel of window is equal to 0, and brightness of the at least one other pixel within the same window is equal to 1. This process is described by the following Boolean function:

$$f\begin{bmatrix} x_1 & x_2 & x_3 \\ x_4 & x_5 & x_6 \\ x_7 & x_8 & x_9 \end{bmatrix} = \bar{x}_5 \& (x_1 \vee x_2 \vee x_3 \vee x_4 \vee x_6 \vee x_7 \vee x_8 \vee x_9). \qquad (5.2.2)$$

Evidently, the function (5.2.2) detects edge if, and only if $x_5 = 0$, and at least one of the other variables is equal to 1, so, it is really the edge detection corresponding to the downward brightness jumps. The following examples illustrate, what is doing with the function (5.2.2).

The edge will be detected by the function (5.2.2) e.g., for the following input values:

$$f\begin{bmatrix} 1 & 1 & 1 \\ 1 & 0 & 1 \\ 1 & 1 & 1 \end{bmatrix} = 1; \quad f\begin{bmatrix} 1 & 1 & 1 \\ 1 & 0 & 1 \\ 1 & 1 & 0 \end{bmatrix} = 1; \quad f\begin{bmatrix} 1 & 0 & 1 \\ 1 & 0 & 1 \\ 0 & 1 & 1 \end{bmatrix} = 1 \quad .$$

The edge will not be detected by the function (5.2.2) e.g., for the following input values:

$$f\begin{bmatrix} 1 & 1 & 1 \\ 1 & 1 & 1 \\ 1 & 1 & 1 \end{bmatrix} = 0; \quad f\begin{bmatrix} 0 & 0 & 0 \\ 0 & 0 & 0 \\ 0 & 0 & 0 \end{bmatrix} = 0; \quad f\begin{bmatrix} 0 & 0 & 0 \\ 1 & 1 & 1 \\ 0 & 0 & 0 \end{bmatrix} = 0 \quad .$$

Variant 3. *Joint detection of edges corresponding to upward and downward jumps.*

Evidently, the considered variant of edge detection may be implemented by the function, which is a disjunction of the functions (5.2.1) and (5.2.2):

$$f\begin{bmatrix} x_1 & x_2 & x_3 \\ x_4 & x_5 & x_6 \\ x_7 & x_8 & x_9 \end{bmatrix} =$$
$$= x_5 \,\&\, (x_1 \vee \bar{x}_2 \vee \bar{x}_3 \vee \bar{x}_4 \vee \bar{x}_6 \vee \bar{x}_7 \vee \bar{x}_8 \vee \bar{x}_9) \vee$$
$$\vee\, \bar{x}_5 \,\&\, (x_1 \vee x_2 \vee x_3 \vee x_4 \vee x_6 \vee x_7 \vee x_8 \vee x_9). \tag{5.2.3}$$

It is possible to clarify now, what is the "double" edge detection, which may be a disadvantage e.g., for precise edge detection of the small objects. It is also easy to verify that the functions (5.2.1) - (5.2.3) give different results in general. Let consider the following (4 x 4) fragment of the binary image:

$$\begin{pmatrix} 0 & 0 & 1 & 1 \\ 0 & 1 & 0 & 1 \\ 1 & 0 & 1 & 0 \\ 1 & 0 & 0 & 0 \end{pmatrix}$$

Let apply the functions (5.2.1) - (5.2.3) to the distinguished pixels (in italic) number *12* and *13*. We will obtain respectively, the following results (only the results in mentioned pixels are considered):

$$\begin{pmatrix} * & * & * & * \\ * & 1 & 0 & * \\ * & * & * & * \\ * & * & * & * \end{pmatrix}; \quad \begin{pmatrix} * & * & * & * \\ * & 0 & 1 & * \\ * & * & * & * \\ * & * & * & * \end{pmatrix}; \quad \begin{pmatrix} * & * & * & * \\ * & 1 & 1 & * \\ * & * & * & * \\ * & * & * & * \end{pmatrix} \quad .$$

It is clearly visible that the results obtained by the functions (5.2.1) - (5.2.3) are different, and that the edge corresponding to the change of brightness exactly between mentioned pixels is twice detected by the function (5.2.3).

As it was mentioned above, a problem of edge detection by narrow direction may be solved using the same approach: an analysis of a 3 x 3 local window around the pixel, and description of such an analysis by Boolean function. A problem of edge detection "by narrow direction" is not new. It is important for example, for detection of the small objects, which are oriented in some specific direction. The most classical solution of such a problem has been proposed in [Prewitt (1970)], and then some methods of a two-dimensional discrete differentiation "by direction" have been proposed in [Pratt (1978)]. We will consider here only the edge detection corresponding to upward brightness jumps (variant 1). The functions for detection of the downward and joint jumps may be easily obtained in the same way, as functions (5.2.2) and (5.2.3) have been just obtained (see above). It is easy to obtain the following Boolean functions for edge detection by direction corresponding to upward brightness jumps on the binary image.

Direction West $\leftrightarrow$ East:

$$f\begin{bmatrix} x_1 & x_2 & x_3 \\ x_4 & x_5 & x_6 \\ x_7 & x_8 & x_9 \end{bmatrix} = (x_5 \& x_4) \& (\bar{x}_1 \vee \bar{x}_7) \vee (x_5 \& x_6) \& (\bar{x}_2 \vee \bar{x}_9). \quad (5.2.4)$$

The examples:

The edge will be detected by the function (5.2.4) e.g., for the following input values:

$$f\begin{bmatrix} 1 & 1 & 0 \\ 1 & 1 & 1 \\ 1 & 0 & 0 \end{bmatrix} = 1; \quad f\begin{bmatrix} 0 & 0 & 0 \\ 1 & 1 & 1 \\ 0 & 0 & 0 \end{bmatrix} = 1; \quad f\begin{bmatrix} 1 & 0 & 0 \\ 0 & 1 & 1 \\ 1 & 0 & 0 \end{bmatrix} = 1 .$$

The edge will not be detected by the function (5.2.4) e.g., for the following input values:

$$f\begin{bmatrix} 1 & 1 & 0 \\ 0 & 0 & 0 \\ 1 & 1 & 1 \end{bmatrix} = 0; \quad f\begin{bmatrix} 0 & 0 & 1 \\ 0 & 1 & 0 \\ 1 & 0 & 0 \end{bmatrix} = 0; \quad f\begin{bmatrix} 0 & 1 & 0 \\ 0 & 1 & 0 \\ 0 & 1 & 0 \end{bmatrix} = 0 .$$

Direction North-West $\leftrightarrow$ South-East:

$$f\begin{bmatrix} x_1 & x_2 & x_3 \\ x_4 & x_5 & x_6 \\ x_7 & x_8 & x_9 \end{bmatrix} = (x_5 \& x_1) \& (\bar{x}_2 \vee \bar{x}_4) \vee (x_5 \& x_9) \& (\bar{x}_6 \vee \bar{x}_8) \quad (5.2.5)$$

The edge will be detected by the function (5.2.5) e.g., for the following input values:

$$f\begin{bmatrix} 1 & 0 & 0 \\ 0 & 1 & 0 \\ 0 & 0 & 0 \end{bmatrix} = 1; \quad f\begin{bmatrix} 1 & 0 & 0 \\ 0 & 1 & 1 \\ 0 & 0 & 1 \end{bmatrix} = 1; \quad f\begin{bmatrix} 1 & 0 & 0 \\ 0 & 1 & 0 \\ 0 & 0 & 1 \end{bmatrix} = 1 \quad .$$

The edge will not be detected by the function (5.2.5) e.g., for the following input values:

$$f\begin{bmatrix} 1 & 1 & 0 \\ 0 & 0 & 0 \\ 1 & 1 & 1 \end{bmatrix} = 0; \quad f\begin{bmatrix} 1 & 0 & 1 \\ 1 & 0 & 0 \\ 1 & 0 & 1 \end{bmatrix} = 0; \quad f\begin{bmatrix} 0 & 1 & 0 \\ 0 & 1 & 0 \\ 0 & 1 & 0 \end{bmatrix} = 0.$$

Direction South-West $\leftrightarrow$ North-East

$$f\begin{bmatrix} x_1 & x_2 & x_3 \\ x_4 & x_5 & x_6 \\ x_7 & x_8 & x_9 \end{bmatrix} = (x_5 \& x_7) \& (\bar{x}_4 \vee \bar{x}_8) \vee (x_5 \& x_3) \& (\bar{x}_2 \vee \bar{x}_6) \qquad (5.2.6)$$

The edge will be detected by the function (5.2.6) e.g., for the following input values:

$$f\begin{bmatrix} 0 & 0 & 1 \\ 0 & 1 & 0 \\ 0 & 0 & 0 \end{bmatrix} = 1; \quad f\begin{bmatrix} 0 & 0 & 0 \\ 0 & 1 & 0 \\ 1 & 0 & 0 \end{bmatrix} = 1; \quad f\begin{bmatrix} 0 & 0 & 1 \\ 0 & 1 & 0 \\ 1 & 0 & 0 \end{bmatrix} = 1 \quad .$$

The edge will not be detected by the function (5.2.6) e.g., for the following input values:

$$f\begin{bmatrix} 1 & 1 & 0 \\ 0 & 1 & 0 \\ 0 & 1 & 1 \end{bmatrix} = 0; \quad f\begin{bmatrix} 1 & 0 & 1 \\ 1 & 0 & 0 \\ 1 & 0 & 1 \end{bmatrix} = 0; \quad f\begin{bmatrix} 1 & 1 & 0 \\ 1 & 1 & 0 \\ 0 & 1 & 1 \end{bmatrix} = 0.$$

Direction North $\leftrightarrow$ South

$$f\begin{bmatrix} x_1 & x_2 & x_3 \\ x_4 & x_5 & x_6 \\ x_7 & x_8 & x_9 \end{bmatrix} = (x_5 \& x_2) \& (\bar{x}_1 \vee \bar{x}_3) \vee (x_5 \& x_8) \& (\bar{x}_7 \vee \bar{x}_9). \qquad (5.2.7)$$

The edge will be detected by the function (5.2.7) e.g., for the following input values:

$$f\begin{bmatrix} 0 & 1 & 0 \\ 0 & 1 & 0 \\ 0 & 0 & 0 \end{bmatrix} = 1; \quad f\begin{bmatrix} 0 & 1 & 0 \\ 0 & 1 & 0 \\ 1 & 1 & 0 \end{bmatrix} = 1; \quad f\begin{bmatrix} 0 & 1 & 0 \\ 0 & 1 & 0 \\ 0 & 1 & 0 \end{bmatrix} = 1 \quad .$$

The edge will not be detected by the function (5.2.7) e.g., for the following input values:

$$f\begin{bmatrix} 1 & 0 & 0 \\ 1 & 1 & 0 \\ 1 & 0 & 1 \end{bmatrix} = 0; \quad f\begin{bmatrix} 1 & 0 & 1 \\ 1 & 0 & 0 \\ 1 & 0 & 1 \end{bmatrix} = 0; \quad f\begin{bmatrix} 1 & 0 & 0 \\ 1 & 1 & 0 \\ 1 & 0 & 1 \end{bmatrix} = 0.$$

It is possible to consider also a more narrow directions, e.g., "only North", "only West", etc., but evidently, the corresponding functions may be easily obtained from the functions (5.2.4)-(5.2.7). It should be noted also that the functions (5.2.4)-(5.2.7) depend in each partial case on seven, and not on nine variables. They detect fact that a brightness jump exists at least at the one pixel in the corresponding direction (in comparison with the central pixel of the window).

So, Boolean functions (5.2.1)-(5.2.3) describe the global edge detection on the binary image, the functions (5.2.4)-(5.2.7) describe the edge detection by direction also on the binary image.

Of course, the edge detection on the gray-scale images is much more interesting. To solve this problem objectively, and independently of the structure of partial image, its statistical characteristics, and dynamic range, the same technique that has been just described for the binary images is proposed to be used. But the following important supplement must be added.

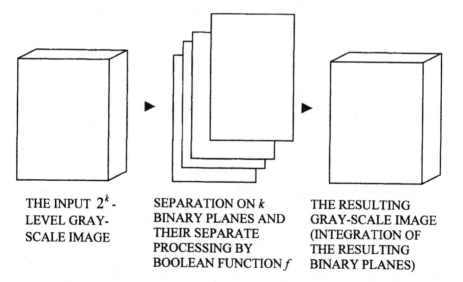

THE INPUT 2^k-LEVEL GRAY-SCALE IMAGE

SEPARATION ON k BINARY PLANES AND THEIR SEPARATE PROCESSING BY BOOLEAN FUNCTION f

THE RESULTING GRAY-SCALE IMAGE (INTEGRATION OF THE RESULTING BINARY PLANES)

Figure 5.2.1. Edge detection on gray-scale images.

A 2^k-level gray-scale image must be separated into k binary planes (don't mix with threshold binarization!), then each binary plane has to be processed separately by one of the functions (2.5.1 - 2.5.7) depending of what kind of edge detection is doing, and the resulting binary images have to be integrated into the resulting gray-scale image. This process is illustrated in Fig. 5.2.1. The integration of the binary planes also may be described by following equation:

$$\widehat{B}_{ij} = \bigoplus_{s=0}^{k-1} [Y_{ij}^s],$$

where $\bigoplus_{s=0}^{k-1}$ denotes direct integration of k binary planes from 0th until $(k-1)^{th}$

to the gray-scale image (color channel); $\widehat{B}_{ij}$ is the resulting integer signal

value in ij^{th} pixel ($\widehat{B}_{ij} \in \{0,1,...,2^k - 1\}$); Y_{ij}^s is the binary value in s^{th} binary

plane of ij^{th} pixel.

It should be emphasized that the separation in binary planes, which we propose here, is absolutely objective operation (in comparison with thresholding binarization, which is strongly subjective because depends on a value of threshold). Direct binarization, which is proposed here, means that the first binary plane is created by bit # 0 of each pixel, the second one by bit # 1, and so on, until k^{th} binary plane (bit # k). The resulting binary planes are integrated into the resulting gray-scale image in the similar natural way: plane number 1 - bit # 0, etc. It will be shown below on the different examples that this approach to edge detection, which is developed here, is much more effective than classical algorithms first of all because of its objective character. It does not depend of any external parameter (e.g., value of threshold), can not "miss" an edge even when difference of brightness in neighbor pixels is equal to 1, does not depend of the characteristics (statistics, dynamic range) of a partial image.

An implementation of the proposed approach should be clarified before we will move to the examples. Of course, the most simple, but not the most effective way is direct evaluation of values of the corresponding Boolean functions (5.2.1) - (5.2.7). It is also possible to use tables of values of these functions, but it also is not effective. As it was mentioned above CNN may be a very nice architecture for the implementation of the considered approach. A lot of the Boolean functions (5.2.1) - (5.2.7) are not threshold, and it is impossible to implement them using traditional CTCNN, or DTCNN. Since the non-threshold functions could be implemented using UBN, it is natural to implement the functions (5.2.1) - (5.2.7) exactly on this neuron. Evidently, the CNN-UBN may be used for the implementation of real-time image processing in such a case.

Let us use one of the learning algorithms described in the Chapter 4 to implement the functions (5.2.1) - (5.2.7) on the UBN. Let us take e.g., the direct learning algorithm for UBN with the rule (4.4.5) - (4.4.6). Application of this learning algorithm to the Boolean functions (5.2.1) - (5.2.7) gives a very quick convergence for all of them. The word "quick" means that the result of learning may be obtained in few seconds even on a very slow (e.g., 33 MHz) computer. The following weighting templates for the CNN-UBN implementation of the functions (5.2.1) - (5.2.7) are the results of learning:

for function (5.2.1), $m=4$ in (3.2.1), 32 iterations:

$$W=(-6.3 \quad -5.6)\begin{bmatrix} (-0.82, 0.32) & (-0.95, -016) & (-0.04, 0.01) \\ (0.25, -1.4) & (-0.32, -0.05) & (-0.03 \ 0.01) \\ (0.0, 0.10) & (0.63, 0.60) & (-0.02 \ 0.01) \end{bmatrix}; \qquad (5.2.8)$$

for function (5.2.2), $m=4$ in (3.2.1), 12 iterations:

$$W=(-1.7 \quad -1.5)\begin{bmatrix} (-0.91, 0.74) & (-0.78, -0.99) & (-0.1, -0.04) \\ (-0.06, -0.08) & (1.4, 1.6) & (-0.86 \ -0.96) \\ (-0.15, -0.07) & (-0.06, -0.05) & (-0.142 \ -0.03) \end{bmatrix}; \qquad (5.2.9)$$

for function (5.2.3); $m=4$ in (3.2.1), 16 iterations:

$$W=(-0.55 \quad -0.82)\begin{bmatrix} (13.0, -25.0) & (0.8, 6.4) & (0.7, 2.9) \\ (1.2, 4.6) & (-1.3, 1.3) & (-2.0 \ 2.0) \\ (-2.9, 1.3) & (-1.9, 4.4) & (1.0 \ 4.3) \end{bmatrix}; \qquad (5.2.10)$$

for function (5.2.4); $m=10$ in (3.2.1), 203 iterations:

$$W=(-8.5, 0.64)\begin{bmatrix} (-0.65, -0.48) & (0.0, 0.0) & (-0.99, 1.6) \\ (1.6, -1.3) & (-2.6, 2.7) & (-0.64, 0.40) \\ (-0.49, -0.39) & (0.0, 0.0) & (-0.09 \ -0.11) \end{bmatrix}; \qquad (5.2.11)$$

for function (5.2.5); $m=22$ in (3.2.1), 19 iterations:

$$W=(1.4 \quad 2.3)\begin{bmatrix} (2.2, -1.5) & (-0.51, 002) & (0.0, 0.0) \\ (-0.39, -0.04) & (0.97, 0.61) & (-0.06 \ 0.3) \\ (0.0, 0.0) & (-0.07, -0.18) & (0.13 \ 0.56) \end{bmatrix}; \qquad (5.2.12)$$

for function (5.2.6); $m=12$ in (3.2.1), 90 iterations:

$$W=(0.64 \quad 6.3) \begin{bmatrix} (0.0, \ 0.0) & (\ 0.01, \ 1.52) & (0.67, \ 1.0) \\ (-0.12, \ 0.26) & (1.04, \ 1.69) & (0.02 \quad 1.4) \\ (0.40, \ 0.29) & (-0.17, \ 0.26) & (0.0 \quad 0.0) \end{bmatrix} ; \qquad (5.2.13)$$

for function (5.2.7); $m=12$ in (3.2.1), 116 iterations:

$$W=(-3.7 \quad 4.8) \begin{bmatrix} (-0.2, \ 0.18) & (-1.4, \ 0.82) & (-0.99, \ 1.6) \\ (\ 0.0, \ 0.0) & (-2.0, \ 0.73) & (0.0 \quad 0.0) \\ (0.06, \ 0.28) & (-0.73, \ 0.31) & (0.04 \quad 0.37) \end{bmatrix} . \qquad (5.2.14)$$

The random numbers from the interval [0, 1] have been chosen as starting weights for the learning process for all functions. Values of m (number of sectors in the UBN activation function (3.2.1)) for all functions are the minimal values that ensure convergence of the learning algorithm for corresponding function.

The templates (5.2.8) - (5.2.14) implement respectively, the edge detection functions (5.2.1) - (5.2.7) on the CNN-UBN. The obtained templates are the universal mean for solution of edge detection problem. It is followed from the objective character of the design the functions (5.2.1) - (5.2.7). They are universal for any partial image. Thus, robustness of the weights is ensured.

Let illustrate the approach to edge detection, which has been just proposed by examples obtained on the CNN-UBN software simulator. Taking into account properties of human vision all edged images are shown here with inverted (negative) palette for more convincing presentation. The example of edge detection on binary image is presented in Fig. 5.2.2. It is clearly visible that all the existing edges are detected despite a complicate form of the analyzed object.

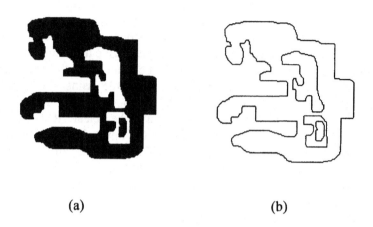

(a)   (b)

Processing by template (5.2.8), function (5.2.1). (a) - the original image;
(b) - the resulting image (upward edges)

Figure 5.2.2. Edge detection on binary image.

Other illustrations show the results obtained for gray-scale images, and show the results obtained for the same images by classical algorithms for comparison. The following important conclusions follow from the example from Fig. 5.2.3. It shows advantages of the approach considered here in comparison with one of the most popular classical methods. Results of the global joint edge detection (function (5.2.3) and template (5.2.10)) are more close to the results obtained by Sobel operator. Simultaneous detection of the upward and downward brightness jumps amplifies the edges corresponding to the great jumps, but edges corresponding to the small jumps are not so well visible. This disadvantage is clearly visible in image corresponding to Sobel operator. A shadow under Lena's does not have a very good contrast in the original, and it is not detected. At the same time a separate detection of upward and downward edges makes possible to obtain a precise detection corresponding even to the smallest brightness jumps. The same shadow is clearly visible in images obtained by processing with Boolean functions. It is also clearly visible even in the image, which is result of disjunctive edge detector (function (5.2.3)), which is more close to the Sobel operator. The reader can compare our result with the results of edge detection on the same image using thresholding, and following application of DTCNN, or CTCNN templates (see e.g., [Chua & Yang (1988)], [Nossek (1996)]). In such a case the result strongly depends of the threshold value, and could not be recognize as objective because some edges (including e.g., edge of the same shadow) are missed. It should be noted that contrast amplification on the edged image obtained by function (5.2.1), or (5.2.2) may be useful for extraction of details. It should be also mentioned that other classical operators like Prewitt, Laplace, conforming Laplace give the result similar to Sobel operator. The same conclusions should be made from the example in Fig. 5.2.4. All shadows on the picture are visible using the methods presented here, also a cloud in the sky. Prewitt operator has not detected these details. From this example it should be clear that separate consideration of the edges corresponding to upward and downward brightness jumps is needed. The corresponding results are different. This important feature is confirmed also by example in Fig. 5.2.5, which presents the results of processing the SAR image of Chernobyl region[4], place of nuclear crash in 1986.

It is clearly visible that quite different objects have been detected on the Fig. 5.2.5 (b) and (c). This is a general feature of the SAR images: to obtain the most objective observation of edges the functions of upward and downward edges detection should be applied to the input image separately.

It is again clear that quality of Sobel edge detection is not satisfactory in comparison with the methods presented here. The following two examples are also taken from practice. Edge detection on angiogramms may be quite important for a correct diagnosis of different vessels pathologies. It is clearly visible that methods, which are described here (Fig. 5.2.6 (b), (c), (d), (e)) are much more effective for detection of blood stream and thrombs in vessels in comparison with classical algorithms. Fig. 5.2.7 presents the results of the edge detection on the electronic microscope image (division of the biological cell). Comparison with classical approaches again shows advantages of the approach presented here: edges of all details, which are even absolutely invisible on the original image have been detected using templates (5.2.8) - (5.2.10) (Boolean functions (5.2.1) - (5.2.3)), but they have not been detected with classical algorithms.

The original image has been made available for experiments by the Ukrainian Center of Air-Space investigation of Earth of the Ukrainian National Academy of Sciences (Kiev, Ukraine)

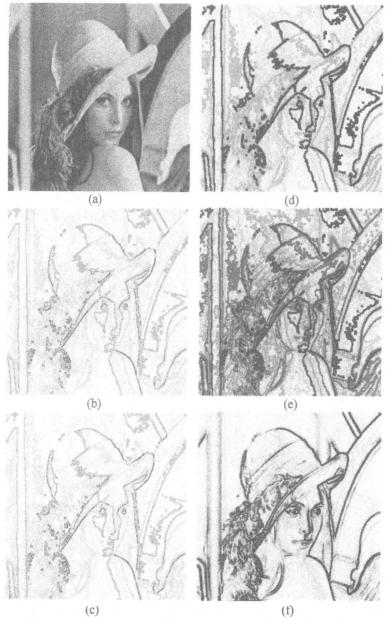

(a) - the original image; (b) - processing by template (5.2.8), function (5.2.1) (upward edges); (c) - processing by template (5.2.9), function (5.2.2) (downward edges); (d) - processing by template (5.2.10), function (5.2.3) (global joint edges); (e) - image from the Fig. (c) corresponding to downward edges with amplified contrast; (f) - processing by Sobel operator.

Figure 5.2.3. Edge detection on "Lena" test image

(a) - the original image; (b) - processing by template (5.2.8), function (5.2.1) (upward edges); (c) - processing by template (5.2.9), function (5.2.2) (downward edges); (d) - processing by template (5.2.10), function (5.2.3); (e) - image from the Fig. (c) corresponding to downward edges with amplified contrast; (f) - processing by Prewitt operator.

Figure. 5.2.4. Edge detection on "Kremlin" test image

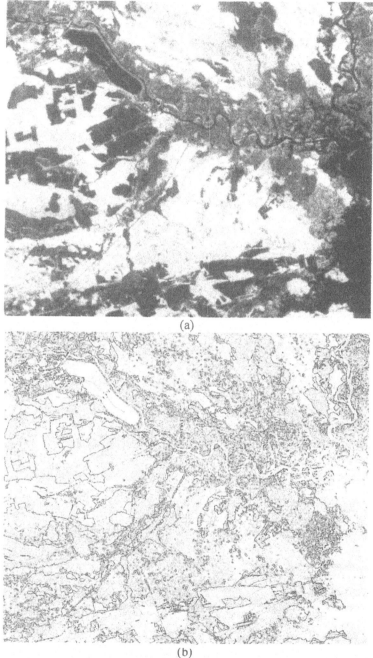

(a)

(b)

(a) - the original image; (b) - processing by template (5.2.8), function
(5.2.1) (upward edges);
Figure 5.2.5. Edge detection on image "Chernobyl"

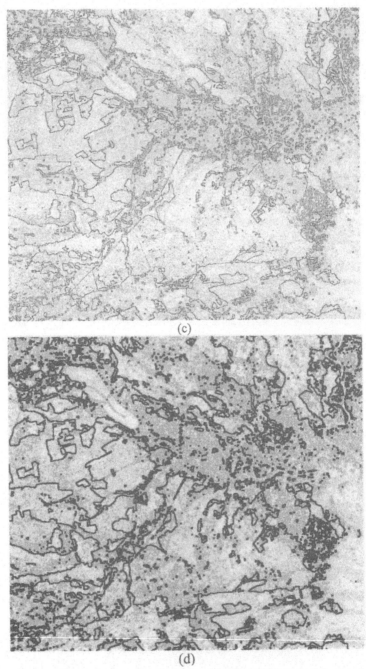

(c)

(d)

(c) - processing by template (5.2.9), function (5.2.2) downward edges);
(d) - processing by template (5.2.10), function (5.2.3) (global joint edges);
Figure 5.2.5. Edge detection on image "Chernobyl"

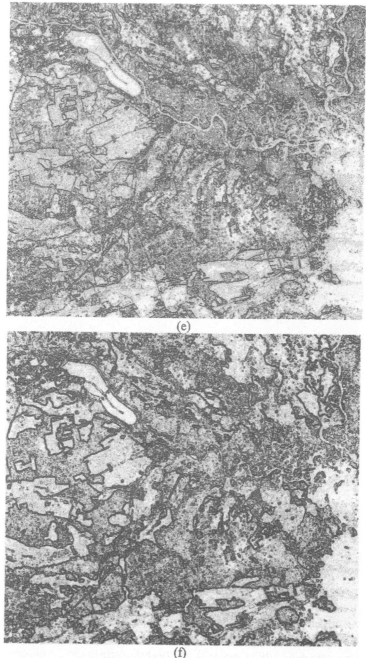

(e) - image from the Fig. (b) corresponding to upward edges with amplified contrast; (f) - processing by Prewitt operator.

Figure 5.2.5. Edge detection on image "Chernobyl"

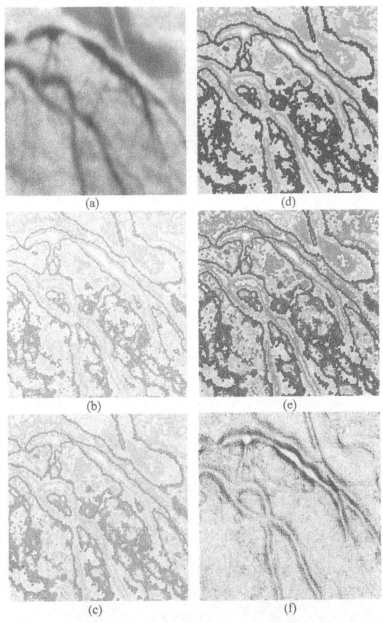

(a) - the original image; (b) - processing by template (5.2.8), function (5.2.1) (upward edges); (c) - processing by template (5.2.9), function (5.2.2) (downward edges); (d) - processing by template (5.2.10), function (5.2.3) (global joint edges); (e) - image from the Fig. (c) with amplified contrast; (f) - processing by Prewitt operator.

Figure 5.2.6. Edge detection on angiogram

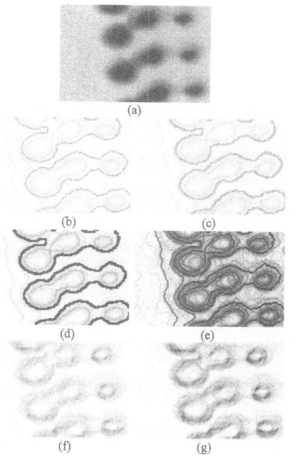

(a) - the original image; (b) - processing by template (5.2.8), function (5.2.1) (upward edges); (c) - processing by template (5.2.9), function (5.2.2) (downward edges); (d) - processing by template (5.2.10), function (5.2.3) (global joint edges); (e) - image from the Fig. (c) corresponding to downward edges with amplified contrast; (f) - processing by Sobel operator; (g)-processing by Prewitt operator.

Figure 5.2.7. Edge detection on electronic microscope image

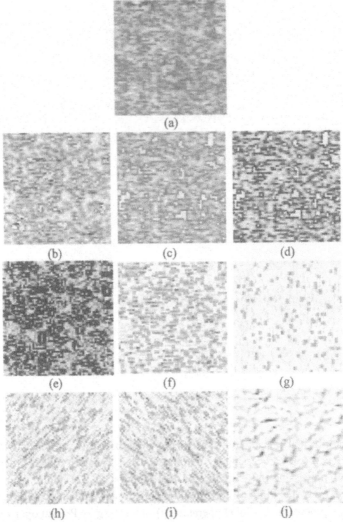

(a) - the original image; (b) - processing by template (5.2.8), function (5.2.1) (upward edges); (c) - processing by template (5.2.9), function (5.2.2) (downward edges); (d) - image from the (c) corresponding to downward edges with amplified contrast; (e)- processing by Sobel operator; (f) - processing by template (5.2.11), function (5.2.4) (West↔East); (g) - processing by template (5.2.14), function (5.2.7) (North ↔ South); (h) - processing by template (5.2.12), function (5.2.5) (North-West ↔ South-East); (i) - processing by template (5.2.13), function (5.2.6) (South-West ↔ North-East); (j) - processing by directed Prewitt mask [Prewitt (1970, Pratt (1978)] (South-West ↔ North-East).

Figure 5.2.8. Edge detection on SAR image of sea

An example, which is presented in Fig. 5.2.8 shows a very good effect first of all of edge detection by narrow direction, which is very important e.g., for monitoring of the sea streams. The fact that edges corresponding to upward and downward jumps are different in general is also confirmed by this example. It is clearly visible that there are different streams directed from West to East, from South-West to North-East, from North-West to South-East, but there is no any stream from South to North. Comparison of edge detection by direction using functions (5.2.4) - (5.2.7) and corresponding templates (5.2.11) - (5.2.14) with a method known as "Prewitt masks" [Prewitt (1970), Pratt (1978)] shows advantages of our approach. The examples in Fig. 5.2.8 (i) and Fig. 5.2.8 (j) show that using Prewitt mask it is impossible to obtain precise detection of all the edges, while the method, which is described here, detects even the smallest brightness jumps.

We hope that the considered examples, and the theoretical considerations convince the reader about the efficiency of the new approach to edge detection and its CNN-UBN implementation. It should be noted also that it is possible to detect edges not on all binary planes, but only on some of them. This approach may be used for image segmentation if the original signal values in the binary planes that are excluded from the operation are preserved (see Fig. 5.2.9). This approach for image segmentation has been proposed in [Aizenberg I. et. al. (1999)]. On the other hand a complete removal from the operation of some binary planes makes possible to perform edge detection on the noisy images. Knowing a dynamic range of the noise it is possible to skip binary planes containing a noisy component. The edge detection algorithm may be directly generalized for color image processing using a separate processing of the color channels.

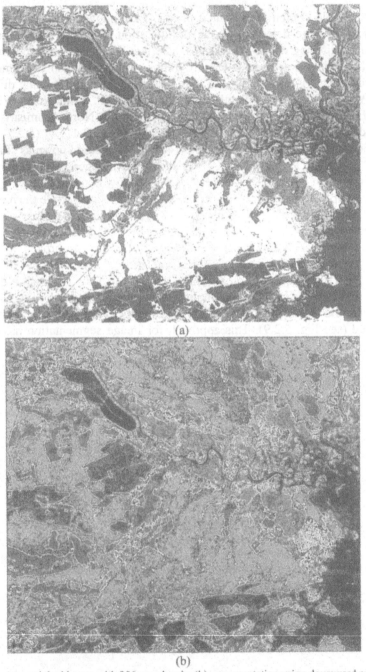

(a) – original image with 256 gray levels; (b) – segmentation using downward edge detection with a preservation of the signal values in 5th and 6th binary planes.

Figure 5.2.9 Edged segmentation of the image "Chernobyl"

Of course, edge detection is not a single application of CNN-UBN. In [Aizenberg N. et. al. (1996a)], [Aizenberg I. (1996)], [Aizenberg I. (1997a] the impulsive noise filtering algorithm based on the same approach that edge detection has been considered. The words "same approach" means an expression of the procedure of impulsive noise detection and filtering through analysis of a 3 x 3 local window around each pixel. For binary images this procedure should be described by Boolean function, and for gray-scale images processing should be organized by separate processing of binary planes. Two Boolean functions have been proposed for detection and filtering of the impulsive noise. One function is oriented on the detection and removal the single impulses (the size of one impulse is equal to one pixel). Another function aims at the detection and removal of more complicated noise, which may be a combination of single impulses and horizontal, vertical, and diagonal "scratches". Both of them are functions of 9 variables. These variables are the values of brightness in a local 3x3 window around its central pixel. The function for detection and filtering of the single impulses is:

$$
f \begin{bmatrix} x_1 & x_2 & x_3 \\ x_4 & x_5 & x_6 \\ x_7 & x_8 & x_9 \end{bmatrix} = \begin{cases} \bar{x}_5, & if\, x_1, x_2, x_3, x_4, x_6, x_7, x_8, x_9 = x_5 \\ x_5, & otherwise \end{cases} \tag{5.2.15}
$$

In other words, if the value of the brightness in the central pixel of a 3x3 local window is not equal to any of other values of brightness of neighbor pixels from this window, function (5.2.15) detects the corresponding pixel as noisy and inverts its brightness. The function (5.2.15) is non-threshold and it cannot be implemented on the single threshold element, but there is no problem with its implementation on the UBN. The following CNN-UBN template has been obtained as a result of the learning (UBN learning algorithm with the rule (4.4.23) - (4.4.24) has been used, the starting weighting vector contained random numbers belonging to the interval [0,1]): $m=18$ in (3.2.1), 85 iterations:

$$
W=(-6.3 \quad -5.6) \begin{bmatrix} (-0.82,\ 0.32) & (-0.95,\ -016) & (-0.04,\ 0.01) \\ (\ 0.25,\ -1.44) & (-0.32,\ -0.05) & (-0.03\ \ 0.01) \\ (0.0,\ 0.12) & (\ 0.63,\ 0.60) & (-0.02\ \ 0.01) \end{bmatrix} \tag{5.2.16}
$$

The function for detection and filtering of combination of single impulses, vertical, horizontal and diagonal "scratches" may be defined by the next formula:

$$f\begin{bmatrix} x_1 & x_2 & x_3 \\ x_4 & x_5 & x_6 \\ x_7 & x_8 & x_9 \end{bmatrix} =$$

$$= \begin{cases} \bar{x}_5, & \text{if lot of } x_1, x_3, x_7, x_9 = \bar{x}_5 \text{ or lot of } x_2, x_4, x_6, x_8 = \bar{x}_5 \\ x_5, & \text{otherwise} \end{cases} \qquad (5.2.17)$$

This is also a non-threshold function, but it can also be implemented on the UBN. The UBN learning algorithm with the rule (4.4.23)-(4.4.24) converges after approximately 8000 iterations (they require few minutes even on a 33 MHz computer) starting from the vector with random components belonging to the interval [0, 1], and gives the next CNN-UBN template for function (5.2.17) ($m = 18$ in (3.2.1)):

$$W=(10.6, 20.6)\begin{bmatrix} (2.40, -1.32) & (-0.92, 0.34) & (0.62, -3.2) \\ (-3.20, -2.51) & (2.32, -0.82) & (0.0, 0.11) \\ (-0.14, 0.12) & (0.02, 0.17) & (-0.06, 0.09) \end{bmatrix} \qquad (5.2.18)$$

We will not stop here on the detailed comparison of the described approach to impulsive noise filtering with other existing and well-known nonlinear filters (median, rank-order, etc.). The reader who are interested can find the corresponding detailed comparisons in [Aizenberg I. (1997a)]. We are far from the statement that our approach is better than other ones. But it has at least one important advantage. Noise removal using the CNN-UBN processing with templates (5.2.16), and (5.2.18) (Boolean functions (5.2.15), and (5.2.17)) is more effective than e.g., median filtering for removal of the noise with a small corruption rate (less than 1 %), and narrow dynamic range (only "salt", or only "peper"). Exactly in such cases image boundaries are almost completely preserved during filtering (they are always smoothed as result of median filtering). An iterative processing usually is requested for the higher corruption rates. It is natural that image boundaries will be smoothed during several iterations, and the final result will be equivalent to the result of median filtering. Knowing a dynamic range of the noise it is possible to process on the CNN-UBN only those binary planes, which contain a noise component. An additional goal is to preserve the image boundaries. Let us consider an example to confirm the declared features of the proposed approach. Fig. 5.2.10 presents the results of impulsive noise reduction using the approach presented here, and using median filter for comparison.

In general, we presented above new class of 2-D nonlinear filters that are reduced to separate processing of the signal binary planes using some Boolean function

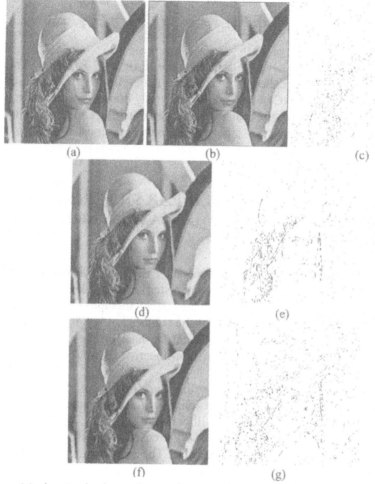

(a) - input noisy image (corruption rate - 0.07, range of the noise [40, 240]); (b) - the results of the processing by template (5.2.18), function (5.2.16); (c) - the difference between input (a) and filtered (b) images: smoothing of the image is minimal; (d) - the results of the third iteration of filtering (three-times recursive processing by template (5.2.16)); (d) -the difference between image (d), and input image (a); (f) - the results of the median filtering of the input image (a) (aperture of the filter is 3x3); (g) - the difference between image (f), and input image (a); It seems that the CNN-UBN approach to filtering proposed above is at least not worse than median filtering.

Figure 5.2.10. Impulsive noise filtering

$$f\begin{bmatrix} x_1 & x_2 & x_3 \\ x_4 & x_5 & x_6 \\ x_7 & x_8 & x_9 \end{bmatrix} = f(x_1, x_2, x_3, x_4, x_5, x_6, x_7, x_8, x_9),$$

and integration of the processing planes into the resulting 2-D object. This family of filters has been called in [Aizenberg I. et. al. (1999)] *Nonlinear Cellular Boolean Filters* (NCBF). Evidently, not only the edge detection and impulsive noise filtering may be reduced to the separate processing of binary planes by Boolean functions, but other filters may be proposed. So it is a nice field for the future work.

3.   MULTI-VALUED NONLINEAR FILTERING AND ITS CNN-MVN IMPLEMENTATION: NOISE REDUCTION

We have to note that we will not present here a global approach to nonlinear filtering. The reader who is interested should address e.g., to [Astola & Kuosmanen (1997)], one of the newest monograph devoted to this problem. A comprehensive observation of the subject is also given in [Pitas & Venetsanopoulos (1990)]. We will concentrate here on the new type of nonlinear filter, which is based on the non-linearity of MVN activation function.

Let us begin from a brief introduction. The purpose of the filtering operation is assumed to be an effective elimination or attenuation of the noise that is corrupting the signal. We will consider here 2-D signals, but of course all considerations may be applied to the case of 1-D signals. Let us consider a N x M image corrupted by a noise. The simplest spatial domain filtering operation is a mean filter defined as

$$\hat{B}_{ij} = \frac{1}{(2m+1)(2n+1)} \sum_{\substack{i-n \leq k \leq i+n \\ j-m \leq l \leq j+m}} B_{kl} , \qquad (5.3.1)$$

where B_{kl} are the values of the input signal, $\hat{B}_{ij}$ is the corrected value of the ij^{th} pixel. The range of indices k and l defines the filter window. The mean filter (5.3.1) is the simplest linear filter. A more effective linear filter is the simple low-pass filter, which may be considered as a generalization of the mean filter (5.3.1). Such a filter is defined like follows:

$$\hat{B}_{ij} = w_0 + \sum_{\substack{i-n \leq k \leq i+n \\ j-m \leq l \leq j+m}} w_{kl} B_{kl} , \qquad (5.3.2)$$

where $w_0 \geq 0$, $w_{kl} > 0$. If $w_0 = 0$, and $w_{kl} = \dfrac{1}{(2n+1)(2m+1)}$ then filter

(5.3.2) is transformed to the filter (5.3.1). It is evident that both filters (5.3.1) and (5.3.2) may be implemented using traditional CTCNN. Such an implementation has been considered in [Aizenberg I (1996)], [Aizenberg I. (1997a)]. For instance, the following CTCNN templates have been proposed for Gaussian noise reduction using filter (5.3.2):

$$A=0;\ B=\begin{bmatrix} 0.11 & 0.11 & 0.11 \\ 0.11 & 0.11 & 0.11 \\ 0.11 & 0.11 & 0.11; \end{bmatrix};\ I=0;\quad A=0;\ B=\begin{bmatrix} 0.1 & 0.1 & 0.1 \\ 0.1 & 0.5 & 0.1 \\ 0.1 & 0.1 & 0.1 \end{bmatrix};\ I=0;$$

$$A=0;\ B=\begin{bmatrix} 0.0625 & 0.125 & 0.0625 \\ 0.125 & 0.25 & 0.125 \\ 0.0625 & 0.125 & 0.0625 \end{bmatrix};\ I=0.$$

The computational simplicity of the linear filters (5.3.1), and (5.3.2) makes them very attractive. But often it is impossible to obtain nice results with them. They reduce noise, but not completely, and simultaneously they smooth the signal.

Maybe the simplest nonlinear filter, which is more complicated from the computational point of view, but usually more effective for the noise reduction is a simple median filter, which is defined as

$$\hat{B}_{ij} = \underset{\substack{i-n \leq k \leq i+n \\ j-m \leq l \leq j+m}}{MED}\ B_{kl}\ , \tag{5.3.3}$$

where *MED* defines a median value within the n x m filter window.

A fundamental difference between the behavior of linear and nonlinear filters may be observed on the comparison of mean and median filters applied to impulsive noise reduction. In the case of a single impulse, the mean filter spreads the impulse but also reduces the amplitude, whereas the median filter totally eliminates the impulse. In fact the median filter with a window size $(2n+1)(2m+1)$ completely removes impulses of length below $(n+1)(m+1)$ independently of their sign [Pratt (1978)]. Thus, a fundamental difference between the behavior of linear and nonlinear filters is the following. The impulse response (that is, the output signal when the input signal is equal to 1 at time instant 0 and zero otherwise) of a nontrivial time-invariant linear filter clearly cannot be a zero-valued sequence [Astola & Kuosmanen (1997)]. It means that the linear filter always smooth a signal. Nonlinear filter may preserve the signal carefully, and usually it is more effective for noise removal despite that it can be idle only for the partial signals.

Different type of nonlinear filter are elaborated and investigated. For example, the order-statistic filters [Sun & Venetsanopoulos (1988)] (also called *L*-filters) are a good mean for removal of the Gaussian and uniform noise. The rank-order filters [Bovik A.C. et al. (1983)] may be used for removal of the mixed (Gaussian or uniform with impulse) noise. The reader who is interested may address also to the weighted median filters [Justusson (1981)], stack filters [Astola & Kuosmanen (1997)], etc.

We would like to use here the specific complex non-linearity of the MVN activation function *P* defined by (2.2.2) to design a new filters family.

It is easy to see from the formulas (5.3.1) - (5.3.3) that a problem of image filtering in the spatial domain is reduced to the replacement of the current value of brightness B_{ij} in ij^{th} pixel by the value $\hat{B}_{ij}$, which is better from some point of view.

Let us consider the linear filters (5.3.1) and (5.3.2). It is well-known that the arithmetic mean is the most popular, but not single mean. Other widely used mean exist, e.g., geometric, harmonic, L_p. But these means is connected with non-linear operations. A general form of nonlinear mean filters has been introduced in [Pitas & Venetsanopoulos (1996)]. It is (for 1-D case for simplicity):

$$B_i = g^{-1}\left(\frac{\sum\limits_{k=-n}^{n} w_k g(B_k)}{\sum\limits_{k=-n}^{n} w_k}\right),$$

where $g(x)$ is the following function $g: R^+ \to R$:

$$g(x) = \begin{cases} x & \text{the arithmetic mean} \\ 1/x & \text{the harmonic mean} \\ \ln x & \text{the geometric mean} \\ x^p, p \in R \setminus \{-1, 0, 1\} & \text{the } L_p \text{ mean.} \end{cases}$$

Since function $g(x)$ is used for averaging of the signal, a following question is natural: is it possible to define another nonlinear mean, and therefore, another type of averaging, which may be used for filtering?

Let $0 \le B \le k-1$ be the dynamic range of a 2-D signal. Let consider the set of the kth power roots of unity. We can put the root of unity

$$\varepsilon^B = \exp(i2\pi B/k) = Y \qquad (5.3.4)$$

to the correspondence with a real value B. Thus, we have the univalent mapping $B \leftrightarrow \varepsilon^B$, where ε is a primitive k^{th} root of unity. One may compare the equations (5.3.4) and (2.2.1), and make a conclusion that they are equivalent.

A two-dimensional *Multi-Valued Filter* (MVF) is a filter, which is defined by the following equality [Aizenberg I. (1997 b]:

$$\hat{B}_{ij} = P(w_0 + \sum_{\substack{i-n \le k \le i+n \\ j-m \le l \le j+m}} w_{kl} Y_{kl}),\qquad (5.3.5)$$

where P is an activation function of multi-valued neuron (2.2.2) with integer output:

$$P(z)=j, \ \ if \ \ 2\pi\,(j+1)/k > arg(z) \ge 2\pi j/k,\qquad (5.3.6)$$

Y_{kl} is obtained from B_{kl} by equation (5.3.4), i,j are the coordinates of the filtered pixel, $n \times m$ is a filter window, w_{kl} are the weighting coefficients (complex-valued in general). Evidently, the equality (5.3.5) defines a class of filters, each of them is defined by the corresponding weighting coefficients.

Let compare the filter (5.3.5) with the linear filters (5.3.1), and (5.3.2). We can make the following conclusion: without the nonlinear function P and complex addition (if Y will be replaced by B), and with $w_{kl} > 0$ filter (5.3.5) becomes the low-pass linear filter in the spatial domain. But filter (5.3.5) is principally nonlinear. First of all the function P is nonlinear, but additionally the complex addition together with the transformation (5.3.4) defines a specific nonlinear averaging of the signal.

A detailed mathematical investigation of the properties of non-linearity carrying in by complex addition and function P may be a subject of a special work. It should be mentioned here that impulse response of MVF is a zero-valued sequence, and this feature brings MVF together with other nonlinear filters. We would like to show that filter (5.3.5) is a very effective method for the reduction of different kinds of noise.

Before we will move to the CNN-MVN implementation of MVF and to the examples of noise reduction, one remark should be given. New different families of nonlinear filters may be considered, if well-known nonlinear spatial domain filters will be linked with MVF. We mean the use of the signal transformation defined by (5.3.4), the complex addition, and application of the function P to the obtained weighted sum. Let consider e.g., a filter, also defined by equation (5.3.5), but simultaneously connected with the weighted rank-order filter. Order-statistic and weighted rank-order filters are based on the compromise between a pure nonlinear operation (ordering), and a pure linear operation (weighting). Actually e.g., a weighted rank-order filter may be defined by the same equation (5.3.2) as a low-pass filter. In such a case conditions under sign Σ have to satisfy the rank-order criteria (order-statistic of addends should be in the limited neighborhood of the order-statistic of the filtered pixel). A weighted rank-order filter, but with the transformation (5.3.4), complex addition instead of real one and the function P applied to the weighted sum, becomes a *rank-order multi-valued filter*

(*ROMVF*). In the same way MVF may be connected with other nonlinear filters.

It is evident from the equations (5.3.5), which defines MVF, and on the other hand (5.1.5), which describes a signal performance by CNN-MVN cell, that a two-dimensional MVF may be implemented using CNN-MVN.

The problem of the MVF implementation using CNN-MVN is reduced to the derivation of templates, by which the corresponding filters may be realized. A solution of such a problem by learning is attractive. At the same time exactly here the corresponding templates may be designed from the heuristic, but natural point of view that our goal is to perform a light smoothing of the signal. A design of templates for the traditional CCNN and DTCNN is very well worked out (see e.g., [Chua & Yang (1988)], [Nossek (1996)], and many problems are solved exactly using this method, not learning. Moreover, the first useful templates for the CCNN and DTCNN have been derived exactly by design in analogy to the known image processing algorithms [Nossek (1996)]. Thus, it will be interesting to obtain some useful templates for the CNN with complex-valued neurons "by design" also in distinction of the previous templates obtained "by learning".

The most simple, but very effective template for reduction of Gaussian, uniform and speckle noise, also as of combination of Gaussian or uniform noise with the impulsive one is the following

$$w_0 = C; \ W = \begin{pmatrix} 1 & 1 & 1 \\ 1 & w_{22} & 1 \\ 1 & 1 & 1 \end{pmatrix}, \tag{5.3.7}$$

where C is a constant (may be equal to zero), w_{22} is a parameter. The template (5.3.7) supposes that the filter (5.3.5) has a 3x3 window. It is designed from the following considerations. If all $w_{kl} = 1$ then a maximal smoothing of the signal will be gotten It is clear that if $w_{22} > 1$ then a smoothing of the signal will be lighter, and the image boundaries will be preserved with more accuracy. At the same time it is clear that for $w_{22} > 10$ the filter will be degenerated because it will not change a signal. To remove Gaussian, uniform or speckle noise, which is not mixed with the impulsive one our recommendation for w_{22} is: $1 \le W_{22} \le 10$. Thus, the CNN-MVN with a 3x3 local connections is sufficient for implementation of the filter (5.3.5) by template (5.3.7).

Let consider some examples. The first one shows a reduction of zero-mean Gaussian noise with a small dispersion, and it is presented in Fig. 5.3.1. The second one shows a reduction of zero-mean Gaussian noise with a large dispersion, and it is presented in Fig. 5.3.2. The third one shows a reduction of mixed zero-mean Gaussian noise with large dispersion, and impulsive noise. It is presented in Fig. 5.3.3. In all examples the results of rank-order filtering of the same images are given for comparison. The Tables 5.3.1 - 5.3.3 contain the objective digital characteristics of the original, corrupted, and filtered images. For all examples MVF and ROMVF preserve image boundaries effectively and ensure more effective reduction of the noise in comparison with rank-order filters.

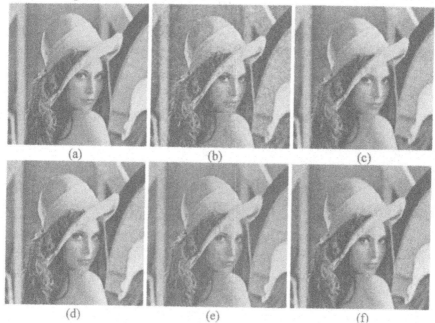

(a) - the original image; (b) - the original image corrupted by zero-mean Gaussian noise with dispersion equal to 0.3 of the signal dispersion; (c) - the results of multi-valued filtering using CNN-MVN (template (5.3.7), W_{22} =2); (d) - the results of multi-valued filtering using CNN-MVN (template (5.3.7), W_{22} =4); (e) - the results of rank-order multi-valued filtering, a filter window 3 x 3, rank-order is equal to 7); (f) - the results of rank-order filtering (a filter window 3 x 3, rank-order is equal to 7).

Figure 5.3.1. Gaussian noise reduction - "Lena"

Table 5.3.1. Digital characteristics of the example given in Fig. 5.3.1

Image	Lena	Lena Noisy	ROF	ROMVF	MVF $w_{22} = 2$	MVF $w_{22} = 4$
ND (σ)	62.1	97.2	12.7	7.1	8.2	10.2
PSNR	57.0	42.2	49.5	53.7	50.4	50.1
SD	0	13.1	12.2	8.7	11.1	10.1

ND(σ) – estimate of the noise dispersion; PSNR – estimate of the peak signal to noise ratio; SD – standard deviation from the original image

(a) - the original image; (b) - the original image corrupted by zero-mean Gaussian noise with dispersion equal to 0.75 of the signal dispersion; (c) - the results of multi-valued filtering using CNN-MVN (template (5.3.7), $W_{22}=2$); (d) - the results of multi-valued filtering using CNN-MVN (template (5.3.7), $W_{22}=4$); (e) - the results of rank-order multi-valued filtering, a filter window 5 x 5, rank-order window is equal to 20); (f) - the results of rank-order filtering (a filter window 5 x 5, rank-order window is equal to 20).

Figure 5.3.2. Gaussian noise reduction - "Kremlin"

Table 5.3.2. Digital characteristics of the example given in Fig. 5.3.2

Image	Kremlin	Kremlin Noisy	ROF	ROMVF	MVF $w_{22} = 2$	MVF $w_{22} = 4$
ND (σ)	81.3	212.6	20.2	6.7	27.1	22.5
PSNR	65.0	29.3	44.2	49.1	46.5	48.9
SD	0	18.7	18.9	14.5	16.3	14.2

ND(σ) – estimate of the noise dispersion; PSNR – estimate of the peak signal to noise ratio; SD – standard deviation from the original image

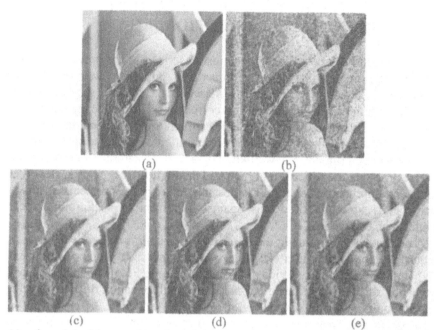

(a) - the original image; (b) - the original image corrupted by zero-mean Gaussian noise with dispersion equal to 0.75 of the signal dispersion, and impulsive noise with corruption rate 10%; (c) - the results of multi-valued filtering using CNN-MVN (template (5.3.7), $W_{22}=1$); (d) - the results of rank-order multi-valued filtering, a filter window 3 x 3, rank-order is equal to 7); (e) - the results of rank-order filtering (a filter window 3 x 3, rank-order is equal to 7).

Figure 5.3.3. Gaussian noise + impulsive noise reduction - "Lena"

Table 5.3.3. Digital characteristics of the example given in Fig. 5.3.3

Image	Lena	Lena Noisy	ROF	ROMVF	MVF
ND (σ)	62.1	491.8	15.1	8.2	12.0
PSNR	57.0	35.3	45.3	48.8	46.2
SD	0	24.6	16.7	13.9	14.3

ND(σ) – estimate of the noise dispersion; PSNR – estimate of the peak signal to noise ratio; SD – standard deviation from the original image

4.   MULTI-VALUED NONLINEAR FILTERING AND ITS CNN-MVN IMPLEMENTATION: FREQUENCY CORRECTION

Multi-valued filters that have been just described in the previous section are not only good for the noise reduction. As was shown recently in [Aizenberg I. & Vandewalle (1997)], [Aizenberg I. (1998)] they are also a very effective mean for the frequency correction. The frequency correction problem may be considered as a specific "anti-filtering". Indeed, the filtering is a reduction of the high frequency. Concentration of the noise spectra in the high frequency domain is a well-known fact. It is natural to consider the inverse problem: amplification of high and medium frequency. It is clear that finding an appropriate solution it is possible to solve the problems of image sharpening, extraction of details against a preventing background, local contrast enhancement.

A natural way for solution the frequency correction problem is filtering in frequency domain. But it is clear that such an approach is complicate from the computational point of view. Direct and inverse Fourier transformation, evaluation of the appropriate filter-mask and multiplication of the spectrum with it require many operations and therefore a lot of time. A very effective approach is the approximation of the corresponding filters in frequency domain by simple linear filters in spatial domain. This approach has been proposed in [Belikova (1990)]. A main idea of this paper is consideration of the extraction, localization, and detection of important features on the preventing image background as a problem of localization of objects or details of a given type on the image. Such a type has been defined as sizes of objects. The idea proposed in [Belikova (1990)] has been developed in [Aizenberg N. et. al. (1994)], where the solution of the problem using traditional CTCNN also has been proposed. A deep consideration of the CTCNN implementation of this approach, and its application to solution of the applied image processing problems has been given in [Aizenberg I. (1997a)].

Let consider this approach in more details.

The global frequency correction implements extraction of the details of medium sizes (in comparison with the image sizes) against a complex image background. Two of the possible filters for solution of such a problem have been proposed in [Belikova (1990)]. These filters are based on the criteria of the minimal mean-square (*ms*) error and the maximum of the signal/noise (*sn*) ratio. The frequency characteristics of these filters are the following.

For the first filter:

$$H_{ms}(i,j) = <|\alpha_\rho(i,j)|^2>_\rho / K_x(i,j) \qquad (5.4.1)$$

$$H_{ms}(i, j)=\|\alpha_{\rho_0}(i, j)\|^2/K_x(i, j) \tag{5.4.2}$$

where $K_x(i, j)$ is the estimate of the power spectrum of the image, α_ρ is the spectrum of the signal of the object, which is in the beginning of the coordinates, $<\ldots>_\rho$ is the meaning by variations of the parameters of the object, and α_{ρ_0} is the spectrum of the object with mean values of the parameters ρ_0.

For the second filter:

$$H_{sn}(i, j)= <\overline{a}_r(i, j)>_r/K_\Phi(i, j) \tag{5.4.3}$$

where $\overline{\alpha}_\rho(i, j)$ is complex-conjugated to $\alpha_\rho(i, j)$, and $K_\Phi(i, j)$ is the estimate of the power spectrum of the background image;

$$H_{sn}(i, j) = \overline{\alpha}_{\rho_0}(i, j)/K_x(i, j), \tag{5.4.4}$$

where $\overline{\alpha}_{\rho_0}(i, j)$ is the complex-conjugated spectrum of the object with mean values of the parameters ρ_0, and $K_x(i, j)$ is the estimate of the power spectrum of the image.

Implementation of the considered filters through orthogonal transformations is a complicate computing problem. Even the FFT algorithm requires $2 N \log(N)$ complex additions and multiplications also as N^2 complex multiplications are required for correction of the spectrum. Using the approximation of the considered filters by two-dimensional filters in the spatial domain the number of operations may be reduced to N^2 multiplications and additions. Moreover, such an approximation may be implemented using CTNN. Taking into account that the filters (5.4.1)-(5.4.4) are used for extraction of the medium spatial frequencies it is possible to approximate them by the following filter [Belikova (1990)]:

$$\hat{B}_{ij}=G_1 B_{ij}+G_2(B_{ij}-B_m)+G_3 B_m+c \tag{5.4.5}$$

where B_m is the local mean value in a window around pixel B_{ij}; B_{ij}, and $\hat{B}_{ij}$ are the signal values in ij^{th} pixel respectively, before and after processing; G_1 and G_3 are the coefficients that define correction of the low spatial frequencies and G_2 defines correction of the high spatial frequencies, c is the mean value of the background after processing. For images with k gray levels (k gradations of each color in color images) recommended values of the weights G are [Aizenberg N. et. al. (1994)]:

$$G_1 \in [0,1], \; G_2 \in [0,10], \; G_3 \in [0,1], \; c \in [0, k-1].$$

The window of the filter (5.4.5) has to be approximately equal to the sizes of objects that are expected to be extracted.

Evidently, the filter (5.4.5) is very convenient for implementation using CTNN because it is reduced to the processing in the local window around the current pixel, or to linear convolution with the weighting window. Sizes of the local neighborhood of the CNN have to be equal to the filter window. By simple transformations we can obtain from the formula (5.4.5) the template for its implementation using CTNN [Aizenberg I. (1997a)]:

$$A=0;\ B=\begin{pmatrix} \dfrac{G_3-G_2}{nm} & \cdots & & \cdots & \dfrac{G_3-G_2}{nm} \\ \cdot & & & & \cdot \\ \cdot & \cdots & G_1+G_2+\dfrac{G_3-G_2}{nm} & \cdots & \cdot \\ \cdot & & & & \cdot \\ \dfrac{G_3-G_2}{nm} & & \cdots & & \dfrac{G_3-G_2}{nm} \end{pmatrix};\ I=C \qquad (5.4.6)$$

where $n \times m$ are the sizes of the local window of the CNN, or window of the filter (5.4.5).

Filter (5.4.5) (and therefore template (5.4.6) for its implementation using CCNN) is an effective way for extraction of details of medium and lower sizes. The extraction of the smallest details and the image sharpening may be gotten by amplification of the highest spatial frequencies. It is possible to obtain a spatial filter for correction of the highest spatial frequencies from (5.4.1)-(5.4.4), as an approximation of the corresponding filter in the frequency domain:

$$\hat{B}_{ij}=B_{ij}+G(B_{ij}-B_m)+c, \qquad (5.4.7)$$

where B_m is the local mean value in window around pixel B_{ij}; B_{ij}, and $\hat{B}_{ij}$ are the signal values in ij^{th} pixel before and after processing, respectively; G is the weighting correction coefficient, c is the mean value of the background after processing. By non-complicated transformations we can obtain from the (5.4.7) template for its implementation of such a filter using CTNN [Aizenberg I. (1997a)]:

$$A=0; \ B=\begin{pmatrix} -\dfrac{G}{nm} & \cdots & & \cdots & -\dfrac{G}{nm} \\ \cdot & & & & \cdot \\ \cdot & & 1+G-\dfrac{G}{nm} & \cdots & \cdot \\ \cdot & & & & \cdot \\ -\dfrac{G}{nm} & & \cdots & & -\dfrac{G}{nm} \end{pmatrix} ; \ I=C \ , \qquad (5.4.8)$$

where $n \times m$ are sizes of the local CNN window or window of the filter (5.4.8), $G \in [0, 10]$ (with $G>10$ intensification of high frequency involves a noisy effect). Filter (5.4.7) is most effective with a small window (from 3x3 to (7x7)).

Different examples of application of the filters (5.4.5) and (5.4.7) that have been implemented by templates (5.4.6), (5.4.8) on the CTCNN software simulator are available in [Aizenberg I. (1997a)]. We will use some of this examples below to compare them with the results obtained using multi-valued filtering. Despite the fact that the filters (5.4.5), and (5.4.7) solve a problem of frequency correction, they have a disadvantage. Such a disadvantage is followed from the linearity of the filters (5.4.5), and (5.4.7). A problem is that these filters are very sensitive to choice of the weighting coefficients G. A little change of their values follows quite significant changes on the processed image. It is reflected in the motion of the global dynamic range of the processed image to "black", or "white" side. As a result some details are being detected and sharpened, but some other (especially the smallest ones) may be missed. Solution of such a problem can be found by replacement of the filters (5.4.5) and (5.4.7) with some nonlinear filters. Such filters have to be good approximators of the filters (5.4.1) - (5.4.4), and should not be much more complicated than filters (5.4.5) and (5.4.7) from the computing point of view, but at the same time they have to be insensitive to the corresponding parameters. The solution has been proposed in [Aizenberg I. (1997b)], and then developed in [Aizenberg I. & Vandewalle (1997)], [Aizenberg I. (1998)].

The multi-valued nonlinear filter (5.3.5) has been obtained as a generalization of the linear filters (5.3.1), and (5.3.2). It is evident that both filters (5.4.5) and (5.4.7) may be obtained from the linear filter (5.3.2) using the corresponding values of the weighting coefficients. Thus, our idea is to obtain the multi-valued filters for frequency correction by generalization of the filters (5.4.5) and (5.4.7) using the same way, as has been used for multi-valued generalization of the filter (5.3.2). Using such considerations we obtain the following generalization of the filter (5.4.5):

$$\hat{B}_{ij} = P\left[c + (G_1 + G_2)Y_{ij} + (G_3 - G_2) \sum_{Y(k,l) \in R_{ij}} Y_{kl}\right], \quad (5.4.9)$$

where Y_{kl} is obtained from B_{kl} by equation (5.3.4), R_{ij} is a local window around pixel Y_{ij}; B_{ij} (Y_{ij}) and $\hat{B}_{ij}$ are signal values in ij^{th} pixel before and after processing, respectively; G_1, G_3 are the coefficients, which define correction of the low frequencies; G_2 defines correction of the high frequencies, c is the constant.

Taking into account (5.3.4) and (5.3.5), filter (5.3.7) is transformed to the following multi-valued filter:

$$\hat{B}_{ij} = P\left[c + (1+G)Y_{ij} - G \sum_{Y_{kl} \in R_{ij}} Y_{kl}\right], \quad (5.4.10)$$

where Y_{kl} is obtained from B_{kl} by equation (5.3.4); R_{ij} is a local window around pixel Y_{ij}; B_{ij} (Y_{ij}) and $\hat{B}_{ij}$ are signal values in ij^{th} pixel before and after processing respectively; G defines correction of the high frequencies, c is a constant.

It is easy to design the templates for a CNN-MVN implementation of the filters (5.4.9) and (5.4.10). A filter for global frequency correction, which is defined by the equation (5.4.9), may be implemented with CNN-MVN using the following template

$$w_0 = c; \quad W = \begin{pmatrix} G_3 - G_2 & \cdots & & \cdots & G_3 - G_2 \\ & \cdot & & & \cdot \\ & & \cdots & G_1 + G_3 \cdots & \\ & \cdot & & & \cdot \\ G_3 - G_2 & & \cdots & & G_3 - G_2 \end{pmatrix}. \quad (5.4.11)$$

To amplify the medium frequencies, and to extract the details of size $n \times m$ the filter window has to be equal exactly to $n \times m$. Thus, the matrix W has to be of the same dimensions. Estimates for the values of the weighting coefficients in (5.4.9) and (5.4.11) are the following. $\frac{nm}{2} \leq G_2 < nm$, $G_2 - 1 < G_3 < G_2$, $\frac{nm}{2} - G_3 < G_1 < 2nm$, where $n \times m$ is a filter window. The reasons for a choice of the values G_1, G_2, G_3 are the same as for the linear filters (5.4.5) and (5.4.7): to ensure a local contrast enhancement within the chosen window it is necessary to move apart a local dynamic range of the image. The lower bounds for the coefficients are chosen experimentally, the upper bounds are chosen from the consideration that higher values leads do the degeneration of the filter. The lower values of the weighting coefficients G_1, G_3 correspond to the stronger amplification of the low frequency. The lower values of the coefficient G_2 correspond to the stronger amplification of the high frequency.

A filter for high frequency correction, which is defined by the equation (5.4.10), may be realized by the following 3 x 3 template ((3 x 3) is the most appropriate window for this filter):

$$w_0 = c; \quad W = \begin{pmatrix} -G_2 & \ldots & \ldots & -G_2 \\ \cdot & & & \cdot \\ \cdot & \ldots & G_1 \ldots & \cdot \\ \cdot & & & \cdot \\ -G_2 & \ldots & & -G_2 \end{pmatrix}, \tag{5.4.12}$$

where c is the constant (may be equal to zero). Estimates for the values of the weighting coefficients in (5.4.10) and (5.4.12) are the following: $\frac{nm}{2} \leq G_1 < 2nm$, $0 < G_2 < 1$. The lower values of G_1 corresponds to the stronger amplification of the high frequencies. It is clear that the filter is degenerated with $G_1 > 2nm$.

It should be noted that: 1) a weighted sum in (5.4.9) and (5.4.10) may be equal to zero (with probability very close to zero) but the function P is not defined on zero. It is possible to use the weight w_0 (see (5.4.11) and (5.4.12)) to correct this situation; 2) k (in equation (2.2.2)) may be equal to the number of gray-levels levels or gradations of each color in color images, but may be little bit higher from the point of view to preserve of the "black"-"white" ("0"-"k-1") inversion (it is not absolutely necessary, because probability of such an inversion is also very close to zero).

The filters (5.4.9), (5.4.10) and their CNN-MVN implementation with the templates respectively, (5.4.11), (5.4.12) are a very good way to solve the frequency correction problem, and give more effective results than their linear predecessors (5.4.5) and (5.4.7). Let us consider several examples.

The results of high frequency correction are presented in the Fig. 5.4.1 - 5.4.3. The first and the second examples show that even a high quality image may be sharpened, and the smallest details may be clearly visible (bricks, from which tower is built, and leafs on the trees in the Fig. 5.4.1(b), hair, eyes in the Fig. 5.4.2(b)). The third example shows that high frequency correction is very effective for sharpening of the small-detailed images and extraction of the smallest details.

The results of global frequency correction are presented in the Fig. 5.4.4 and Fig. 5.4.5. It is interesting that linear filter (5.4.5) can't detect the microcalcifications (the small light points, which are signs of the breast cancer) on the image from the Fig. 5.4.4 (a). But the problem of their detection is solved using multi-valued filter.

(a)   (b)

(a) - the original image; (b) correction by the MVF
(filter (5.4.10), template (5.4.12), G_1 =8.5)

Figure 5.4.1. High frequency correction using MVF - "Kremlin"

(a)   (b)   (c)

(a) - the original image; (b) - correction using MVF (filter (5.4.10), template (5.4.12), G_1 =7.5); (c) - correction by linear filter (5.4.7), template (5.4.8), G=1.5 (a global dynamic range of the image has been moved to the "white" side)

Figure 5.4.2. High frequency correction using MVF - "Lena"

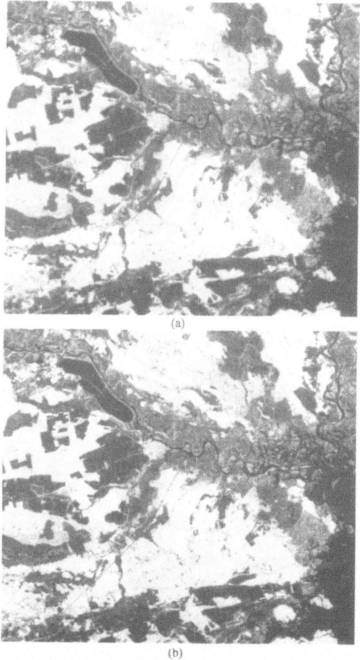

(a)

(b)

(a) - the original image; (b) - correction using MVF (filter (5.4.10), template (5.4.12), $G_1 = 6.0$)

Figure 5.4.3. High frequency correction using MVF - "Chernobyl"

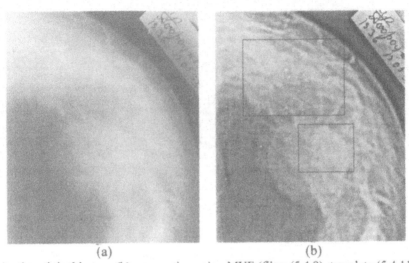

$$(a) \qquad\qquad\qquad (b)$$

(a) - the original image; (b) - correction using MVF (filter (5.4.9), template (5.4.11), window 15 x 15, $G_1 = -1.0; G_2 = 115.0; G_3 = 114.5$). The microcalcifications are being visible after processing (small light spots, especially within marked regions. Identification of the microcalcifications is very important for diagnostic of the cancer).

Figure 5.4.4. Global frequency correction using MVF - "Mammogram"

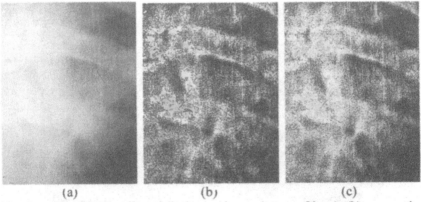

$$(a) \qquad\qquad (b) \qquad\qquad (c)$$

(a) - the original low-quality medical x-ray image (tumor of lung); (b) - correction using MVF (filter (5.4.9), window 25x25, template (5.4.11), $G_1 = -1.0; G_2 = 315.0; G_3 = 314.5$), tumor and its structure is being very visible; (c) - correction by the linear filter (5.4.5), template (5.4.6), $G_1 = 0.4; G_2 = 6.0; G_3 = 0.5$ (a dynamic range has been moved to the "white" side, and some of the important details are still invisible).

Figure 5.4.5. Global frequency correction using MVF - "Tumor of lunq"

Thus, the multi-valued nonlinear filters, which are based on the non-linearity of MVN activation function may be successfully used both for noise reduction, and frequency correction. It is very convenient from our point of view, especially taking into account the effective results of the applications. The most fresh application of MVF together with edge detection using CNN-UBN for ultrasound medical image processing is presented in [Aizenberg I. et. al. (1998c)]. In this paper a general strategy of low quality image processing using CNN is developed (such a strategy has been considered in [Aizenberg I. (1997a)], but at that time multi-valued filters were not carried out yet). In summary we can take the following strategy of operations on images:

- 1) If the processed image is corrupted by the "white" noise or a combination of the "white" and impulsive noise, first such an image must be processed using CNN-MVN by the filter (5.3.5), template (5.3.7);
- 2) If the image is corrupted by the impulsive noise it must be processed using CNN-UBN by one of the templates (5.2.16) or (5.2.18) (filters (5.2.15) or (5.2.17)).
- 3) For correction of the blurring (smoothing) the filtered image must be processed using CNN-MVN by template (5.4.11) or (5.4.12) (filters (5.4.9) or (5.4.10)). Templates (5.4.11) and (5.4.12) may also be used for amplification of the medium and high spatial frequencies and thus for extraction of the image details.
- 4) A highly effective operation for detection of the small details is precise edge detection defined by the functions (5.2.1)-(5.2.7)) and implemented using CNN -UBN by the templates respectively (5.2.8)-(5.2.14).
- 5) Combination of the images, obtained in steps 3 and 4 (or 2 and 4), is a highly effective means for sharpening of the image - all details including smallest will be extracted by this operation.

We will return to the multi-valued filtering in the Section 6.3, where its implementation for solution of the super resolution problem will be presented.

5.   CONCLUSIONS

Cellular neural networks with universal binary and multi-valued neurons were considered in this Chapter. It has been shown that CNN-UBN and CNN-MVN complement classical CTCNN and DTCNN. CNN-UBN and CNN-MVN are very effective for the implementation respectively, of cellular neural Boolean filters and multi-valued filters. The first family of filters is used for precise edge detection on binary and gray-scale images and for impulsive noise filtering. It was proposed to solve problem of precise edge detection by description of edge-detection using Boolean function and its implementation using CNN-UBN. A problem of impulsive noise filtering has been solved in the same way. It was shown that detection of edges corresponding to upward, downward and disjunctive (upward, downward, or upward and downward together) jumps of brightness must be considered separately and should be described by the different Boolean functions.

The second family of filters (multi-valued ones) may be applied to noise removal and frequency correction (extraction of image details). Templates for the implementation of the considered filters using CNN-UBN and CNN-MVN have been obtained.

Chapter 6

Other Applications of MVN and MVN-based Neural Networks

Different applications of MVN and MVN-based neural networks are presented in this Chapter. Models of the associative memory that are based on the different neural networks with MVN and are considered. A single-layered MVN based neural network for image recognition is considered. An approach to pattern recognition based on orthogonal spectra analysis is developed. The application of MVN for time series prediction is presented. Finally a super resolution problem is solved using prediction of the spectra on MVN and using the iterative preliminary approximation of the spectra, and its final approximation with multi-valued filters.

1.   MODELS OF THE ASSOCIATIVE MEMORY

One of the natural applications of neural networks is a pattern recognition. It is based on the possibility to train a network so that it can restore, recognize, or predict a process described precisely (or approximately) by a mapping. One of the most popular neural recognition systems is associative memory or content addressable memory. Its main feature is a possibility to restore a pattern using only its fragment or to restore a corrupted pattern using its uncorrupted part. A problem has been formulated and deeply studied by T.Kohonen (see [Kohonen (1977), (1884)]. Since there are many different kinds of neural networks, the different authors returned to this problem and proposed different solutions. May be the most popular proposal was made in the famous paper [Hopfield (1982)]. We will not compare here the different associative memories. The interested reader can find such a comparison e.g., in [Haykin (1984)], [Hassoun (1985)]. Our aim is to show that it is possible to design effective associative memory using MVN. Moreover, the first application of MVNs was exactly an associative memory based on CNN-MVN [Aizenberg & Aizenberg (1992)].

The fully connected neural network used as binary associative memory in [Hopfield (1982)] was a good background for further work in two directions: generalization of such an associative memory for the multi-valued case and its simplification taking into account that a fully connected network is a complicated structure. The following extensions of the Hopfield associative memories to multi-valued case should be mentioned among other. In [Kohring (1992)] it was proposed to assign each bit of multi-valued state representation to different neurons in a cluster. Then the corresponding bits can be processed on separate Hopfield networks independently from each other. But such a network is too complicated, and very sensitive to errors because change of one bit leads to impossibility of the restoring of pattern. Another approach has been proposed in [Zurada et. al. (1994)], where a Hopfield network with $(n+1)$-valued neurons has been considered. An

activation function of these neurons was $f(h, n) = 2/n \sum_{i=1}^{n} f(h - \Theta_i) - 1$

with n inflection points Θ_i. It was shown that such a network has an ability to remove distortions from noisy gray-scale images. But learning algorithm for such a network was not very fast, also as full connectivity of the Hopfield network is a technical problem. In [Tan et. al. (1990)] CNN with threshold neurons and Hebb learning rule was proposed to be used as associative memory for binary images. Actually CNN are very attractive because of their local connections feature. It is their great advantage in comparison with fully connected network. Exactly this paper was a background for development the associative memory based on CNN-MVN. As it was mentioned above such a network was proposed to be used as associative memory in [Aizenberg & Aizenberg (1992)].

Let consider CNN-MVN of a dimension $N \times M$ with $n \times m$ local connections. Our goal is to use it as associative memory and to store there the gray-scale images $P_i, \ldots, P_s$ of a size $N \times M$. The natural questions are: what is the capacity of such a memory (what is an estimate for the number s)? What about the restoration of the corrupted images? What about the learning of such a network?

To store the gray-scale images with l gray levels MVN with activation function (2.2.2), and $k \ge l$ should be used as a basic CNN cells. $k > l$ should be chosen to prevent a possible inversion "black" to "white" or back. E.g., for $l=256$ it is enough to take $k=264$ (see Fig. 2.2.1). In such a case the rest of the sectors in the complex plane will not be used. If the weighted sum fall into one of these sectors then the sectors 256-259 should correspond to the output value ε^{255}, and sectors 260-263 should correspond to the output ε^{0}.

Each cell of the CNN-MVN, who is used as associative memory, implements the following correspondence:

$$P_m^{ij} = f(x_1, \ldots, x_{(nm+1)/2}, \ldots, x_r), \; m = 1, \ldots, s; \; r = nm, \qquad (6.1.1)$$

if a cell with the coordinates ij has a feedback connection. Here P_m^{ij} is a brightness value of image number m in the ij^{th} pixel. But evidently, if each cell has a feedback connection then $x_{(nm+1)/2} = P_m^{ij}$, and this means that the mapping, which we have to implement on the each cell is the following:

$$x_{(nm+1)/2} = f(x_1, \ldots, x_{(nm+1)/2}, \ldots, x_r), \; m = 1, \ldots, s; \; r = nm. \qquad (6.1.2)$$

The function (6.1.2) is always a multiple-valued threshold function because it may be always implemented with the weighting vector, for which $W_{(nm+1)/2} = 1$, and all the other weights are zero-valued. Such a weighting vector is trivial, and could not be used for restoration of the information in the associative memory. But it is principally important that the function (6.1.1) (or (6.1.2)) is always a multiple-valued threshold function. To find a non trivial weighting vector the learning algorithm for MVN should be used. Since a learned function is a multiple-valued threshold function, the learning always converges. To obtain a non-trivial weighting vector we should start the learning process from an arbitrary vector (e.g., from the random numbers taken from the interval [0, 1]). Learning of one (ij^{th}) cell consist of sequential updating of the weights for all images $P_1, \ldots, P_s$ to ensure the implementation of the function (6.1.1). Learning of the network consist of sequential learning of all the cells. A number of images, which may be stored in such a CNN-MVN based associative memory, is huge. Indeed, a potential estimate for s is defined by functionality of each cell, and number of cells. The function (6.1.1) takes k^r values. Since this function always may be implemented using MVN and our network contain $N \times M$ cells, the upper bound for s is NMk^r. Of course, it will be more and more difficult to obtain a non trivial weighting vector for each cell when the number of the stored patterns increases. A problem is that modulo of the weight $w_{(nm+1)/2}$ will become larger and larger in comparison with the other weights. But an estimate NMk^r is so great number even for $k=256$, $r= 9$, $N=M=3$ that it is difficult to say, where it is possible to find so many patterns to store. So, a real problem is not in trivial weights, but in principal limitation of the CNN for restoration of the corrupted signal. It is clear that since CNN is a locally connected network, a corruption of a window with the sizes compatible with $M \times N$ makes a restoration very difficult. A corruption of the window of sizes equal to $M \times N$ or more makes it impossible to restore the information within such a window. It means that actually only the impulse noise may be removed from image stored in the described associative memory. At the same time the impulse noise with the so high corruption rate (until 35-40%) can be completely removed. Of course, more iterations are necessary for removal of the noise with the corruption rate close to the mentioned upper estimate. One iteration means that all network's cells perform the formula (5.1.5) one time. In all our experiments with associative memory models we used the data base, which contains the portraits of forty people (one per person). All images are 64 x 64 with 256 gray levels A fragment of this testing data base is presented in Fig. 6.1.1.

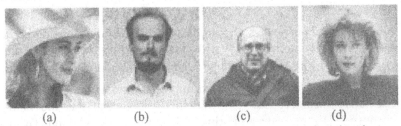

   (a)   (b)   (c)   (d)

Figure 6.1.1 Fragment of the associative memory testing data base

A software simulator of the network has been used. A learning of the 64 x 64 CNN-MVN with 3 x 3 local connections using learning algorithm with the Strategy 2 (see Section 4.1) and the learning rule (4.2.1) is going quickly and requests from 0.5 to 2.5 hours depending on a basic computer. The starting weights for all the neurons were random numbers from the interval [0, 1]. Some experiments with restoration of the images that were corrupted by the impulse noise have been done after learning. The results of one of them are presented in the Fig. 6.1.2.

        (a)   (b)   (c)

(a) image from Fig. 6.1.1 (a) corrupted by impulse noise, corruption rate is 30%;

(b) restoration on CNN-MVN, 50 iterations;

(c) - restoration on CNN-MVN, 150 iterations

Figure 6.1.2. Restoration of the image corrupted by impulse noise in the CNN-MVN associative memory

The image from the Fig. 6.1.2 (c) practically completely coincides with the image (a) from the learning set (see Fig. 6.1.1) (standard deviation is 0.07).

It is possible to consider CNN-MVN without feedback connections. In such a case the function (6.1.1), which describe a mapping implementing by ij^{th} neuron, is transformed to the following function of n-1 variables:

$$P_m^{ij} = f(x_1, ..., x_{(nm+1)/2-1}, x_{(nm+1)/2+1}, ..., x_r),$$

$$m = 1, ..., s; \ r = nm.$$

(6.1.3)

It is impossible to say about this function in advance whether it is threshold or not. Since the learning set is limited, the function (6.1.3) is partially defined. Hence it is likely that one can find such a value of k, for which this function will be threshold, as partially defined function. In such a case a central pixel of each n x m window is removed from the learning process. It contains a positive moment because the weighting vectors for all neurons will be always non trivial. But it considerably increases a number of iterations, and time needed for learning. At the same time the CNN-MVN with the cells performing the correspondence (6.1.3) has approximately the same possibilities that the CNN-MVN performing the correspondence (6.1.1).

Thus, the CNN-MVN model for associative memory is much more simpler than fully connected Hopfield model. It makes possible to store a huge number of patterns, and to remove impulse noise from them. A disadvantage of such a model is impossibility to restore a pattern with a completely corrupted fragment, whose sizes are equal or higher than the sizes of a CNN local window. It is clear that a pattern with a destroyed regular region is also can not be restored.

A natural question followed from the last conclusion: is it possible to develop another locally-connected network, which will be free from the mentioned disadvantage? A positive answer has been given in [Aizenberg N. et. al. (1995), (1996b)].

A network, which we would like to present here, has a local connections feature also as cellular one. Each neuron of this network also is connected with a limited number of other ones. But in cellular network each neuron is connected only with neurons from its nearest neighborhood (Fig. 5.1.1). In network, which will be considered, a function of the connections for each neuron is generated as some random function. An example of such a network is shown in Fig. 6.1.3. The ij^{th} neuron is connected with 8 other neurons and with itself. Numbers of neurons, from which ij^{th} neuron receives the input signals are chosen randomly.

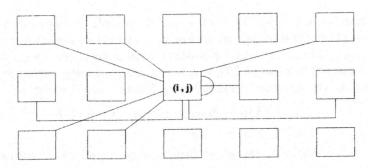

Figure 6.1.3. Fragment of the network with random connections

Let again our goal is to store the gray-scale images P_i, ..., P_s of a size N x M in the associative memory. We will again use a N x M neural network with MVN as basic cells. It will not be a cellular network, but a network with random connections. In other words the number of neurons, whose outputs are connected with the given neuron, are defined by the random numbers generator. It ensures that even if some regular fragment of the stored pattern will be destroyed or corrupted, then the connections with the neurons corresponded to the non corrupted region make possible to restore a corrupted information. Let each neuron of the network have n inputs (it is connected with n-1 other ones and with itself). Therefore each (ij^{th}) neuron performs a mapping described by the following function:

$$P_m^{ij} = f(x_1, ..., x_{n-1}, x_n) = f(x_1, ..., x_{n-1}, P_m^{ij}), \ m = 1, ..., s, \qquad (6.1.4)$$

where P_m^{ij} is a brightness value in the ij^{th} pixel of image number m. Also as the functions (6.1.1) and (6.1.2), the function (6.1.4) is a multiple-valued threshold function. It always may be implemented on the MVN by the trivial weighting vector with $w_n = 1$, and all the other components equal to zero. To obtain a non trivial weighting vector, the learning algorithm should be used also as in the case of CNN-MVN based associative memory. A potential capacity of such a network is defined by the functionality of each neuron, and it is equal to NMk^n. It is also possible to consider a network without feedback connections. In such a case a mapping implemented by each neuron is described by the following multiple-valued function of n-1 variables:

$$P_m^{ij} = f(x_1, ..., x_{n-1}), \ m = 1, ..., s. \qquad (6.1.5)$$

Comparable advantages and disadvantages of both models (with the mappings (6.1.4) and (6.1.5)) are the same that of CNN-MVN models with the mappings respectively (6.1.1) and (6.1.4). So we will not repeat them. A general advantage of network with random connections in comparison with

CNN is the possibility to restore patterns not only with impulse corruption, but with completely destroyed regions.

Let consider the example with the same data base. The learning set consists of forty 64 x 64 images that are portraits of the different people (the fragment of the data base is shown in the Fig. 6.1.1). We used a software simulator of the network, in which each neuron has been connected with 29 other ones, and with itself. Thus, a mapping for each neuron has been described by the function (6.1.4) of 30 variables. The learning algorithm with the Strategy 2 (see Section 4.1), and the learning rule (4.3.1) sequentially has been applied to all neurons of the network. The starting weights was taken as random numbers from the interval [0, 1]. The learning set for each neuron contained the values of brightness in the pixels from all images which had the same coordinates as the given (ij^{th}) neuron. A learning of the network is a sequential learning of all neurons. This learning procedure converges quickly. It requests from 0.2 to 1 hour depending on a basic computer. The results of image restoration are presented in the Fig. 6.1.4 - 6.1.5. In both examples the restored images almost coincide with the original one.

(a)   (b)   (c)

(a) image from Fig. 6.1.1 (a) with 75% fragment completely replaced by uniform noise; (b) restoration on network with random connections, 40 iterations; (c) - restoration on network with random connections, 150 iterations

Figure 6.1.4. Restoration of the image
with the fragment completely replaced by noise
on the neural network with random connections

The standard deviation for image form the Fig. 6.1.4 (c) is equal to 0.05, and for image from Fig. 6.1.5 (c) the standard deviation is equal to 0.17.

The considered two models of associative memory show that fully connectivity of the Hopfield network is not necessary for ensuring of the high capacity of the associative memory and its ability to restore a corrupted signal. It is easy to obtain an estimate for capacity of the fully connected NxM MVN-based network. Since each neuron of such a network implements a mapping described by the multiple-valued threshold function

$$P_m^{ij} = f(x_1, ..., x_{n-1}, x_n) = f(x_1, ..., x_{NM-1}, P_m^{ij}), \ m = 1, ..., s$$

then a top estimate for capacity of such a memory is NMk^{NM}. But NMk^n (top estimate for capacity of the MVN based network with random connections) is also a huge number. It means that knowing a number of patterns, which should be stored in the memory, it is possible to chose a connections structure (number of inputs in each neuron, etc.).

(a) image from Fig. 6.1.1 (a) corrupted by impulse noise, corruption rate is 30%; (b) restoration on network with random connections, 15 iterations; (c) - restoration on network with random connections, 80 iterations

Figure 6.1.5. Restoration of the image corrupted by impulse noise on the neural network with random connections

Of course, a low estimate for capacity of the associative memory based on probability of an error in the network response should be also evaluated. In [Jankovski et. al. (1996)] independently of us a Hopfield network based on MVN, and used as associative memory for gray-scale images has been introduced. It should be noted that a Hebbian learning proposed in the mentioned paper for such a network is not effective, and considerably reduce its possibilities in comparison with our learning algorithm based on the rules (4.2.1) or (4.3.1). But at the same time the method proposed in this paper for obtaining of the low estimate of network capacity may be used here.

Let ε^α be a correct output of ith neuron (we will omit 2-D structure of the network, it is not relevant here). It means that

$$P(w_0 + w_1 x_1 + \ldots + w_n x_n) = P(z) = \varepsilon^\alpha.$$

Let some of the inputs (or even all of them) $x_1, \ldots, x_n$ be corrupted. In such a case the weighted sum may be presented like follows:

$$w_0 + w_1(x_1 + \tilde{x}_1) + \ldots + w_n(x_n + \tilde{x}_n)) = z + \tilde{z}, \qquad (6.1.6)$$

where $\tilde{z} = w_0 + w_1 \tilde{x}_1 + \ldots + w_n \tilde{x}_n$ is a total noise component. To obtain a low estimate of the network capacity we have to obtain a probability that the addend $\tilde{z}$ will move a weighted sum from the desired sector number α to another one. Of course, such a probability has to be obtained for all neurons of the network. Let $L=NM$ be the total number of neurons in the network, $\tilde{z}_{ls}$ is a noise component for l^{th} neuron corresponding to the sth stored pattern, S is a total number of stored patterns. In such a case we obtain the following expression for the total noise component :

$$\tilde{Z} = \sum_{l=1}^{L}\sum_{s=1}^{S}\tilde{z}_{ls} \; .$$

To solve our problem we have to calculate the probability distribution of $\tilde{Z}$. But what is $\tilde{Z}$? Actually it is a sum of LS complex-valued components. Let all $\tilde{z}_{ls} \in \{1, \, \varepsilon, \, \varepsilon^2, ..., \, \varepsilon^{k-1}\}$. Such an assumption is a strong condition, but it simplifier our considerations. In such a case $\tilde{z}_{ls}$ may be considered as independent random k-valued numbers from the distribution

$$\Pr(\tilde{z}_{ls} = \varepsilon^j) = 1/k, \; j = 0, \, 1, \, ..., \, k-1 \qquad (6.1.7)$$

with a mean value of zero and a variance of one. According to the Lindberg-Levy central limit theorem [Feller (1960)], as LS tends to infinity, the distribution for $\tilde{Z}$ becomes a complex Gaussian distribution with mean zero and variance σ^2. Therefore

$$\lim_{LS\to\infty}\Pr(\mathrm{Re}\,\tilde{Z} < \tilde{Z}_x, \; \mathrm{Im}\,\tilde{Z} < \tilde{Z}_y) = \frac{1}{2\pi\sigma^2}\int\limits_{-\infty}^{Z_x}\int\limits_{-\infty}^{Z_y} e^{-u^2/2\sigma^2}e^{-v^2/2\sigma^2}\,dudv,$$

where $\tilde{Z}_x$ and $\tilde{Z}_x$ are the real and imaginary parts of the variable $\tilde{Z}$. Let z_{ls} is a desired value of the weighted sum for the lth neuron and s^{th} pattern. Since the actual output is equal to $z_{ls} + \tilde{z}_{ls} = h_{ls}$, we are able to estimate the distribution for h_{ls} by the following 2-D Gaussian distribution:

$$f_{h_{ls}}(x, y) = \frac{1}{2\pi\sigma^2}e^{-(x-m_x)^2/2\sigma^2}e^{-(y-m_y)^2/2\sigma^2}, \qquad (6.1.8)$$

where x and y are the real and imaginary parts of the variable h_{ls}. Let without loss of generality $P(z_{ls}) = \varepsilon^0 = 1$. In other words the 0th sector is desired. In such a case the probability p that the output of the network is erroneous is equal to probability that $z_{ls} + \tilde{z}_{ls} = h_{ls}$ does not fall to the 0th sector of the complex plane. The last condition is equivalent to the following: $\dfrac{x}{y} > \cot(2\pi/k)$ (see Fig. 2.2.1). The last inequality is transformed to $x > y \cdot \cot(2\pi/k)$. Therefore

$$p = 1 - \Pr\big(\mathrm{Re}\,h_{ls} > \mathrm{Im}\,h_{ls}\cdot\cot(2\pi/k)\big).$$

The probability p may be estimated by integrating the distribution $f_{h_{ls}}(x, y)$ in the region of the 0th sector:

$$p = 1 - 2 \int\limits_{0}^{\infty} \int\limits_{y \cot(2\pi/k)}^{\infty} f_{h_{ls}}(x, y) dx\, dy \ . \tag{6.1.9}$$

Let $\lambda = (x - m_x)/\sqrt{2\sigma}$ and $\mu = (y - m_y)/\sqrt{2\sigma}$. Thus from (6.1.8) and (6.1.9) we finally obtain using the same considerations that presented in [Jankovski et. al. (1996)]

$$p = \frac{1}{2}\left[1 + \int\limits_{0}^{\infty} \frac{2}{\sqrt{\pi}} e^{-\mu^2} erf(\mu \cot(2\pi/k) - \frac{1}{2\sigma}) d\mu \right].$$

Evidently, the capacity of any MVN based associative memory reduces for increasing k and number of patterns, which should be saved. At the same time it is necessary to understand that formal error (shifting of the weighted sum into neighbor sector, or even not neighbor, but no far away from the desired one for some neurons) should not be considered as a fatal error. It is clear that for gray-scale images with gray levels ≥ 256 such errors are not fatal, and the image may be restored and recognized. It is clear from the examples considered above. In all of them there are some incorrectly restored pixels. But at the same time it is absolutely impossible to say that the corresponding images have not been restored. A standard deviation of the restored images from the original ones is very small.

Thus, it is possible to use the different MVN based neural networks as associative memories. They are more effective than Hopfield-like memories based on traditional neurons. The reader can compare our results e.g., with [Ramaher et. al. (1993)], where the experiments similar to our ones were provided on the neurocomputer "Synapse-1" with Hopfield fully connections architecture. A learning set in that work consists only of 8 portrait images (in our case 40, and this number is restricted only by the speed of conventional computer, which we used). It was possible to restore only a small corrupted region (window around eyes). Ratio of sizes of this region to sizes of image is 1:8. In our experiments there were no problems with restoration of the 3/4 of image (example in Fig. 6.1.4). At the same time it is premature to reject that such memories because of their enormous potential. Evidently they fail to restore the rotated image (if an angle of rotation is more than 5 degrees), the shifted image, and, of course, another image corresponding to the same object e.g., another portrait of the same person. To solve such problems a preprocessing of the data is necessary because learning directly in spatial domain is not enough. We will return to this problem in the Section 6.2.

2. IMAGE RECOGNITION USING MVN-BASED NEURAL NETWORK

It is important to note that the restoration of the information in an associative memory is not the same as pattern recognition. A typical problem, which may be solved in the associative memory, is the restoration of the concrete pattern using its fragment. A typical problem of pattern recognition consists in finding the class, to which some object belongs. The MVN based associative memories have been considered in the Section 6.1. Despite their high potential they could not be used e.g, for face recognition, which is very popular image recognition problem. To recognize the different portraits of a same person, shifted or rotated images other approaches should be used.

Solution of the image recognition problems using neural networks became very popular during last years. Many corresponding examples are available (see e.g., the recent publications [Petkov et. al. (1993)], [Foltyniewicz (1996)], [Lawrence et. al. (1997)] . On the other hand many authors consider image recognition reduced to analysis of the some number of orthogonal spectrum coefficients on the different neural networks (see e.g., [Ahmed & Rao (1975)], [Foltyniewicz (1996)]). Here we will present a new approach to image recognition based on the following background: 1) high functionality of the multi-valued neuron and quick convergence of its learning algorithm 2) well-known fact about concentration of the signal energy in the small number of low-frequency spectral coefficients [Ahmed & Rao (1975)]; 3) not so well known, but important fact about concentration of the information about signal in phase of the Fourier spectrum [Aizenberg I. & Yaroslavsky (1989)].

As it was mentioned above a disadvantage of the networks used as associative memories is impossibility of the recognition of shifted or rotated image, also as image with changed dynamic range. To break these disadvantages and to more effective use of MVN features, we would like to propose here a new type of the network, learning strategy and data representation (frequency domain will be used instead of spatial one). We will use here the results, presented in the papers [Aizenberg I. & Vandewalle (1997)], [Aizenberg I. (1998)], [Aizenberg & Aizenberg (1999)] and the results of our recent experiments.

Consider N classes of objects, which are presented by images of n x m pixels. The problem is formulated into the following way: we have to create recognition system based on neural network, which makes possible successful identification of the objects by fast learning on the minimal number of representatives from all classes.

To make our method invariant to the rotations, shifting, and to make possible recognition of other images of the same objects we will move to frequency domain representation of objects. It has been observed (see e.g.,

[Ahmed & Rao (1975)] that objects belonging to the same class must have similar coefficients corresponding to low spectral frequencies. For different classes of discrete signals (with different nature and length from 64 until 512) sets of the lowest 20-60 coefficients are very close one to other for signals from the same class from the point of view of learning and analysis on the neural network [Ahmed & Rao (1975)]. This observation is true for different orthogonal transformations. It should be mentioned that a neural network proposed in [Ahmed & Rao (1975)] for solution of the similar problem was based on the obvious threshold elements. Thus only two classes of objects have been considered. In the terms of neural networks to classify object we have to train a neural network with the learning set contained the spectra of representatives of our classes. Then the weights obtained by learning will be used for classification of unknown objects.

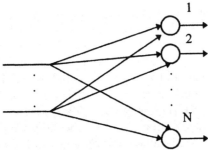

N classes of objects - N neurons

Figure 6.2.1. Neural network for image recognition

We propose the next structure of the MVN based neural network for the solution of our problem. It is a single-layer network, which has to contain the same number of neurons as the number of classes we have to identify (Fig. 6.2.1). Each neuron has to recognize the patterns belonging to its class and to reject any pattern from any other class.

To ensure a more precise representation of the spectral coefficients in the neural network they were normalized, and their new dynamic range after normalization has became $[0, k-1]$. We used two different models for frequency domain representation of our data. The first one is using the low part of Cosine transformation coefficients. The second one is using the phases of the low part of Fourier transformation coefficients. In the last case we used a property of Fourier transformation that the phase contains more information about the signal than the amplitude (this fact is investigated in [Aizenberg I. & Yaroslavsky (1989)]).

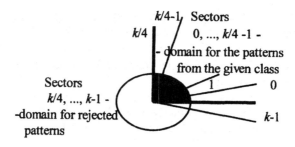

Figure 6.2.2. Reservation of the domains for recognition - 1st model

The best results for the first model were obtained experimentally, when we reserved the first *l*=*k*/4 (from the *k*=512) sectors on the complex plane (see (2.2.2), and Fig. 2.2.1) for accepted patterns. The other 3/4 sectors correspond to rejected patterns (Fig. 6.2.2). The best results for the second model were also obtained experimentally, when we reserved the first *l*=*k*/2 (from the *k*=512) sectors on the complex plane (see (2.2.2), and Fig. 2.2.1) for accepted patterns. The other *k*/2 sectors correspond to rejected patterns (Fig. 6.2.3).

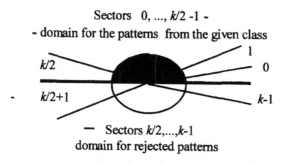

Figure 6.2.3. Reservation of the domains for recognition - 2nd model

Thus, for both models output values 0, ..., *l*-1 for the ith neuron correspond to classification of object as belonging to i[th] class. Output values *l*, ..., *k*-1 correspond to classification of object as rejected for the given neuron and class respectively. Hence the three results of recognition during the training: 1) output of the neuron number *i* belongs to {0, ..., *l*-1} (it means that network classified pattern as belonging to class number *i*); outputs of all other neurons belong to {*l*, ..., *k*-1}; 2) outputs of all neurons belong to {*l*, ..., *k*-1}; 3) outputs of the several neurons belong to   {0,

..., *l*-1}. Case 1 corresponds to exact (or wrong) recognition. Case 2 means that a new class of objects has been appeared or to non sufficient learning or not representative learning set. Case 3 means that number of neurons' inputs is small or inverse, is large, or that learning has been provided on the not representative learning set.

The proposed structure of the MVN-based network and approach to solve of the recognition problem have been evaluated on the example of face recognition. Experiments have been performed on the software simulator of neural network. We used MIT faces data base [Turk & Petland (1991)], which was supplemented by some images from the data base used in our experiments with associative memories (see Section 6.1). So our testing data base contained 64 x 64 portraits of 20 people (27 images per person with different dynamic range, conditions of light, situation in field). So, our task was training of the neural network to recognize twenty classes. Fragment of the data base is presented in Fig.6.2.4 (each class is presented by the single image within this fragment).

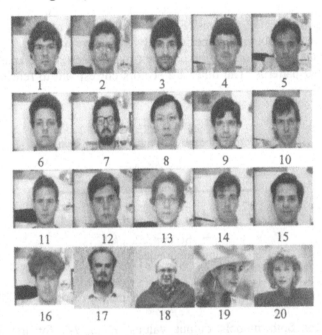

Figure 6.2.4. Fragment of the testing data base for face recognition (1 image per each class)

According to the structure of the network proposed above, our single-layer network contains twenty MVNs (the same number, as number of classes). For each neuron we have the following learning set: 16 images from the class corresponding to given neuron and 2 images for each other class (so 38 images from other classes). Let us describe the results obtained for both models.

Model 1 (Cosine transformation).

According to the scheme presented in Fig. 6.2.2 sectors 0, ..., 127 have been reserved for classification of the image as belonging to the current class, and sectors 128, ..., 511 have been reserved for classification of the images from other classes. The learning algorithm for MVN with the strategy 1 (see Section 4.1) and the learning rule (4.3.1) have been used. So, for each neuron $q=63$ for patterns from the current class and $q=319$ for other patterns in the learning rule (4.3.1).

The best results have been obtained for 20 inputs of the network or for 20 spectral coefficients, which are inputs of the network. More exactly, there are 20 low coefficients (from second until sixth diagonals, zero-frequency coefficient has not been used). Choice of the spectral coefficients from the diagonals of spectrum is based on the property of 2-D frequency ordered spectra: each diagonal contains the coefficients corresponding to the same 2-D frequency ("zigzag", see Fig. 6.2.5).

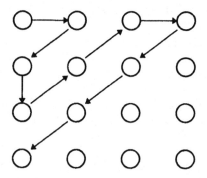

Figure 6.2.5. Choice of the spectral coefficients, which are inputs of neural network

We have got quick convergence of the learning for all neurons. Computing time of the software simulator is about 3 - 15 seconds per neuron[5], which corresponds to 2000-3000 iterations. It is necessary to make an important remark: if it is impossible to obtain convergence of learning for the given k it is necessary to change it and to repeat the process.

For testing, twelve images for each person, which were not presented in the learning set, and are other or corrupted photos of the same people, have been shown to the neural network for recognition. For classes 1, 2, and 17 testing images are presented respectively, in Fig. 6.2.6 - 6.2.8. Results are the following. The number of incorrectly identified images for all classes

---

[5] Depending on the type of a basic computer

(neurons) is from 0 (for 15 classes from 20) to 2 (8%), except classes No 2 and 13. For both classes No 2 and 13 this number is increased to 3-4. May be, it is influence of the same background, on which photos have been made, and very similar glasses of both persons (see Fig. 6.2.4). To improve the results of recognition in such a case the learning set should be expanded. From our point of view it is not a problem because additional learning is very simple. On the other hand, increasing of the number of classes, which have to be identified, also is not a problem. It is always possible to add necessary number of neurons to the network (Fig. 6.2.1) and to repeat learning process beginning from the previous weighting vectors.

Model 2 (Fourier transformation).

The results corresponding to model 2 are better. According to the scheme presented in Fig. 6.2.3 sectors 0, ..., 255 have been reserved for classification of the image, as belonging to the current class, and sectors 256, ..., 511 have been reserved for classification of the images from other classes. The learning algorithm for MVN with the strategy 1 (see Section 4.1) and the learning rule (4.3.1) have been used again. So we chose $q=127$ for patterns from the current class and $q=383$ for other patterns in the learning rule (4.3.1).

The results of recognition sequentially improved with increasing of number of network inputs. It should be noted that such a property was not noticed in method 1. The results of recognition remain the same for more than 20 coefficients.

The best results have been obtained for 405 inputs of the network (or for 405 spectral coefficients, which are inputs of the network) and from this number all the results remain the same. The phase of the spectral coefficients has been chosen again according to the "zigzag" rule (Fig. 6.2.5).

For almost all classes 100% successful recognition has been obtained. For classes "2" and "13" 2 images have been swapped, but this mistake has been easily corrected by additional learning. The reasons of this mistake are clearly visible. In fact both images have the same background, and both persons wear very similar glasses.

In order to compare both methods, and to estimate the precision obtained by learning, we consider in Table 6.2.1 the numbers of the sectors (from 512), where the weighted sum is located for images from the class No 17.

The considered examples show both the efficiency of the proposed solution of image recognition problem and excellent potential of MVN.

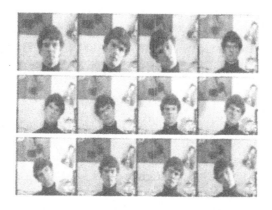

Class "1": 100% successful recognition

Figure 6.2.6

Class "2", model 1: 9 out of 12 images (75 %) are recognized. Incorrectly recognized images are marked by "*"

Class "2", model 2: 100% successful recognition

Figure 6.2.7

Class "17" - 100% successful recognition

Figure 6.2.8

Table 6.2.1. Number of sectors, where the weighted sum is located for recognition of the images presented in Fig. 6.2.8 (class 17).

Image	1	2	3	4	5	6	7	8	9	10	11	12
Method 1, Sector (board ers are 0,127)	60	62	62	102	40	34	65	45	99	65	35	46
Method 2, Sector (board ers are 0,255)	126	122	130	129	120	135	118	134	151	126	107	119

3.   TIME SERIES PREDICTION

Sections 5.1-5.2 already showed that the high functionality and the fast convergence of the learning algorithm for MVN open many possibilities for applications related to pattern recognition. In the same papers [Aizenberg N.

et. al. (1995), (1996b)], where the associative memory based on the MVN-network with random connections has been proposed, the results of time-series prediction using single (!) MVN have been also described. This problem is not new, of course. Many authors considered the time series prediction using neural networks during last years. E.g., the following references may be given: [Hornik et. al. (1989)], [Barron (1993)], [Asar & McDonald (1994)], [Suykens, Lemmerling et. al. (1996)],)], [Suykens & Vandewalle (1998)]. Neural networks are successfully used in financial forecasting, in forecasting of resources consumption (electricity, gas), etc. A general conclusion from the above list of references is that multi-layer perceptrons are universal approximators in the sense that they can approximate any continuous nonlinear function arbitrarily well on a compact interval. An approach, which we will prospectively present here, is based on use of single MVN, its high functionality and quickly convergent learning algorithm.

Let us take a time series

$$x_0, x_1, x_2, ..., x_{n-1}, x_n, x_{n+1}, ..., x_i, ... \qquad (6.3.1)$$

The range of x is $[a, b]$; $a, b \in R$. We will use the n-input MVN for extrapolation of the series (6.3.1). To create a multiple-valued (k-valued) function for learning and extrapolation, we first of all normalize the data to change the input range $[a, b]$ to the range $[0, k-1]$. Let us assume for simplicity that (6.3.1) represents a series in normalized form. Let us create a function, which will express a correspondence between series components. Such a function has to be learned and then extrapolated to the unknown domain. Let us assume that each x_i depends on the previous n components $x_{i-n}, ..., x_{i-1}$ of the series. Thus we obtain:

$$\left.\begin{aligned} x_n &= f(x_0, x_1, ..., x_{n-1}) \\ x_{n+1} &= f(x_1, x_2, ..., x_n) \\ &\cdots\ \cdots\ \cdots\ \cdots\ \cdots \\ x_i &= f(x_{i-n}, x_{i-n+1}, ..., x_{i-1}) \end{aligned}\right\} \qquad (6.3.2)$$

Let a learning set contains s values: $x_n, x_{n+1}, ..., x_i$, $i=n+s-1$. To represent the function f according to (2.2.1) we have to apply the MVN learning algorithm with the rule (4.2.1) or (4.3.1) to obtain the weights $w_0, w_1, ..., w_n$. The function (6.3.2) is a partially defined multiple-valued function. After finishing the learning, it is possible to extrapolate the values of $x_{i+1}, x_{i+2}, ...$ using formulas (2.2.1), (6.3.2) and the weighting vector, which is a result of the learning process.

It should be noted that using MLP-like network from MVN it is possible to obtain a forecaster similar to [Barron (1993)]. But our goal was to check possibilities exactly of the single MVN to solve a problem of time series prediction. It is clear that these possibilities will be high if an appropriate learning set and appropriate value of n in (6.3.2) may be chosen, also as the learning algorithm for these parameters and data relatively quickly converges. The described approach has been tested on several examples.

Example 6.3.1. Forecasting of the currency exchange rate.

It is a very popular problem. The testing series in our example was series presenting USD/DM exchange rate. Components of the series have been presented according to (6.3.1), and the function (6.3.2) has been defined. x_0 corresponds to the closing exchange rate of New York stock exchange on April 23, 1984. The following values also have been taken as closing results of New York stock exchange after the mentioned date. For the forecasting with a high precision, data has been presented with 4 digits after decimal point. An exchange rate in each current day has been presented as a function of 1036 variables (number of the week days in 4 years) in 1024-valued logic. k=1024 has been taken to ensure a high precision of learning and forecasting. Thus, the input data have been normalized using linear transformation, and their range changed to [0, 1023]. After learning and forecasting the inverse transformation to change a range of data to natural form should be applied. The learning process has been organized according to the Strategy 2 (see Section 4.1) using the learning rule (4.2.1). The starting values of weights have been taken as random numbers from the interval [0, 1]. A convergence of the learning has been gotten after about 100000 iterations for 880 elements in the learning set. The number of iterations is naturally increasing with increasing number of elements in the learning set. For the learning set contained 1495 elements about 1000000 iterations are needed for convergence. It should be mentioned that using integer-valued technique of learning presented in the Section 4.5 such a huge number of iterations may be performed during considerably limited time. E.g., 1000000 iterations requests about 100 hours on Pentium-II-266 MHz computer. The results of prediction corresponding to different cardinality of the learning set are presented in the Fig. 6.3.1.

The horizontal axis shows number of the week days (0 corresponds to the next day after finishing of learning). The vertical axis presents the value of 1 USD in DM. Almost all predicted values are different from actual ones at least in the 4th digit after decimal point. It may be a result of the normalization the input data, and the inverse transformation after prediction. We obtained coincidence of the 10-12 actual and predicted values followed by the last value including to the learning set, for the learning sets with cardinality 880, 981, and 1122 (with a precision until 3[d] digit after decimal point). A coincidence with the same precision for the learning set of cardinality 1495 has been obtained for 90 values. It is impossible to make a global conclusion concerning quality of forecast beginning from the moment when actual and predicted values begin to differ in the 3rd digit after decimal point. The accumulation of error involves more, and more differences in actual and predicted values. A conclusion, which follows from the described experiment is that an increase of the cardinality of the learning set involves an increase of the precision of prediction. This conclusion is confirmed also by the following example.

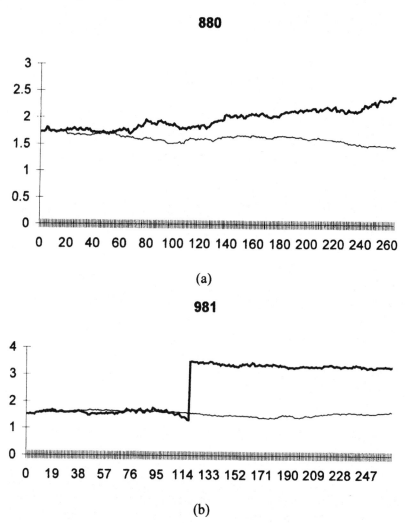

the bold curve - predicted exchange rate, the light curve - actual exchange rate; (a) - the learning set consist of data for 880 days; (b) - the learning set consist of data for 981 days

Figure 6.3.1. USD/DM exchange rate forecasting on the single MVN

1122

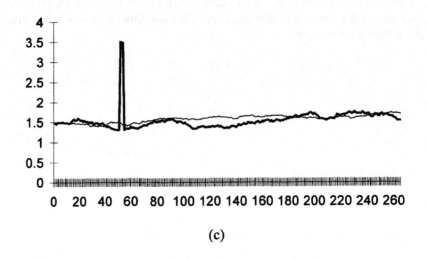

(c)

1495

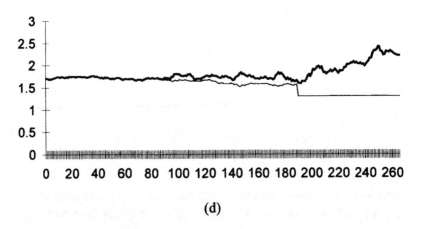

(d)

the bold curve - predicted exchange rate, the light curve - actual exchange rate; (c) - the learning set consist of data for 1122 days; (b) - the learning set consist of data for 1495 days

Figure 6.3.1. USD/DM exchange rate forecasting on the single MVN

Example 6.3.2. Extrapolation of the trigonometric functions.

Fig. 6.3.2 presents the results of prediction of the 259 values of the function

$$\sin(2\pi j / 1024), \; j=573, ..., 832, \qquad (6.3.3)$$

and their comparison with the actual values of the same function. The function (6.3.2) has been taken as a function of 259 variables in 1024-valued logic (k=1024). The results for the learning set of a cardinality 313 are shown in the Fig. 6.3.2. Since 313+259=572, the number of the first extrapolated value is j=573.

It is evident from the Fig. 6.3.2 that the predicted values are almost equal to the desired values after learning on a half of period. Apart horn some very small differences, the behavior of the function SIN is represented by extrapolation with very high precision.

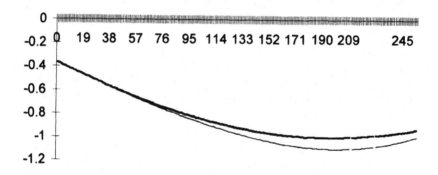

the bold curve - predicted values, the light curve - actual values; on the horizontal axis the numbers of predicted values, and actual values with the same numbers are presented (0 corresponds to the j=573).

Figure 6.3.2. Prediction of the function (6.3.3).

We hope that considered examples illustrate the high potential of multi-valued neurons. We will return again to the time series prediction on the MVN in the Section 6.4. One of the methods of the solution the super resolution problem will be based on the application of the same technique.

4.   SOLUTION OF THE SUPER RESOLUTION PROBLEM USING MULTI-VALUED FILTERING AND MVN-PREDICTION OF THE SPECTRAL COEFFICIENTS

The last application of MVN, which we would like to present here, is the solution of the super resolution problem.

Increasing the resolution is a very important problem, especially for the processing of images with small details. The simplest, but not a good way to solve the problem is interpolation. It is usually possible to get a factor of 2-3 increase of image by bilinear or bicubic interpolation, but this is not a solution of a super resolution problem. The highest frequency part of the spectrum of an interpolated image is very close to constant, so it is impossible to speak about increasing of the resolution.

So the real increase of the resolution may be obtained only by extrapolation of the spectrum to the highest frequency domain.

The problem of extrapolation (restoring) of the spectrum of the spatial-limited object is one of the most important problems of the optics [Stark (1982)]. Any optical system may be considered as a two-dimensional low-frequency filter. Thus, only restricted part of the frequencies participates in the acquisition of the image. It means, that resolution of an optical system may be increased by restoring of the highest frequencies, or, by extrapolation of the spectrum to the highest frequency domain. A couple of different methods have been proposed for solution of the problem. Some of them are beautiful from the point of view of mathematics, but lead to the solution of integral equations (see e.g., [Goodman (1968)], [Huang (1975)]), which is a complicate numerical problem. Solutions which are much more close to practice have been proposed in [Shaker & Steenart (1978)], [Stark (1982)]. These solutions are based on the two fundamental facts [Huang (1975)]: 1) the two-dimensional Fourier-image of a spatial-limited function is an analytical function in frequency domain; 2) if the analytical function in frequency domain is defined exactly on the limited subdomain, it is defined on the all domain, where it is analytical. The iterative procedures, which have been proposed in [Shaker & Steenart (1978)] for 1-D signals, and in [Stark (1982)] for 2-D signals, are directed to the simultaneous restoring of the unknown part of spectrum corresponding to the highest frequencies, and values of the signal in the spatial domain. Computer implementation of both algorithms is not difficult, but they have some disadvantages. The starting zero-values of spectral coefficients and signal values can not be recognized as a good solution. Such a way suppose in advance that the restored part of a spectrum will be rather smooth, and the signal values in the restored domain will not be so close to the ideal values.

Twenty-two and eighteen years passed respectively from publication of [Shaker & Steenart (1978)], and [Stark (1982)]. No really new effective algorithms for solution of the super resolution problem have been discovered. We would like to propose here two solutions of the super resolution problem (see [Aizenberg I. & Vandewalle (1998)], [Aizenberg I. et. al. (1998b)], [Aizenberg I. (1998)]).

The first one is based on the same fundamental approach, as proposed in [Shaker & Steenart (1978)] and [Stark (1982)], but will be free from its disadvantages. It means that we will provide the same iterative procedure, but with the following significant differences: 1) the starting values of the restoring spectrum and signal will not be zero-valued, or constant; 2) a final correction of the spectrum and signal will be realized by the nonlinear multi-valued filtering in spatial domain; 3) not only Fourier, but Cosine and Walsh (ordered by Walsh) transformations will be used.

The second approach to solution of the super resolution problem is based on the possibility of time series prediction using single MVN. Assuming that the high frequency spectral coefficients are the function of lower ones we will predict the unknown highest frequency spectral coefficients.

Let us start with the first approach.

Let $f(x, y)$ be a two-dimensional signal, or discrete image of a $n \times n$ sizes (without loss of generality), which is defined on the spatial subdomain $\widetilde{A} \subset A$. The function

$$F(u, v) = \Phi\big[f(x, y)\big] ; \; u, v \; \in \{0, 1, ..., n-1\} \; ,$$

where Φ is Fourier or cosine or Walsh (ordered by Walsh) transform, is the spectrum of the signal f. The problem is an extrapolation of the function F to the domain $u, v \in \{0, 1, ..., n-1, n, ..., 2n-1\}$. In other words it is evaluation of the values of the signal $f(x, y)$ (or image of $2n \times 2n$ sizes) on the whole domain A. Thus it is always a dual problem: extrapolation of the spectrum and interpolation of the image. Such a duality is illustrated in Fig. 6.4.1. An idea used in [Shaker & Steenart (1978)] and [Stark (1982)] may be formulated like follows: if spectra is extrapolated to the domain $u, v \in \{0, 1, ..., n-1, n, ..., 2n-1\}$ the coefficients corresponding to the subdomain $u, v \in \{0, 1, ..., n-1\}$ should be taken from the original spectra F. Let us suppose that

$$g(x, y) = \begin{cases} f(x, y), \; if \, (x, y) \in \widetilde{A} \\ s(x, y), \; if \, (x, y) \in A \backslash \widetilde{A} \end{cases} , \quad (6.4.1)$$

where $s(x, y)$ is a uniform noise with the same mean value that $f(x, y)$ and a small dispersion[6]. We will use the function g defined by (6.4.1) as a starting

---

[6] It is also possible to take the result of bilinear or bicubic interpolation of the signal $f(x, y)$ as $s(x, y)$ instead of the uniform noise. The final results will be very close each other.

approximation in our iterative and recursive algorithm. Equation (6.4.1) is illustrated in Fig. 6.4.2.

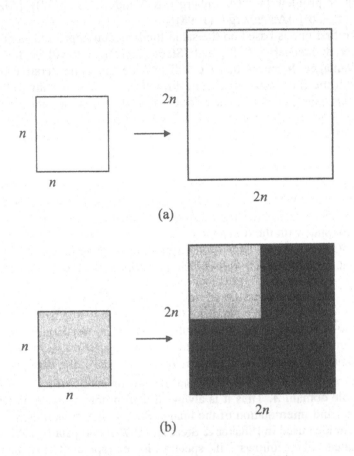

(a) - spatial domain; (b) - frequency domain:
n x n lower coefficients in extrapolated spectra are the same as in original one

Figure 6.4.1. Duality of the spectra extrapolation / image interpolation problem

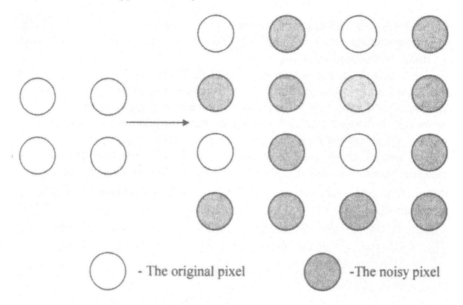

Figure 6.4.2. Illustration for the equation (6.4.1).
Original image 2 x 2, and noisy image 4 x 4.

We can consider $g(x, y)$ as $f(x, y)$, corrupted by the additive uniform noise, moreover, we know in such a case, that our signal is corrupted only within domain $A \setminus \tilde{A}$. It means that our problem may be formulated as a problem of a noise reduction and correction of the highest frequencies. It is the first and very important difference with approach proposed in [Shaker & Steenart (1978)] and [Stark (1982)], where $s(x, y)=0$ was proposed to be used in the representation similar to (6.4.1). We are proposing the following algorithm for the doubling of the resolution.

1. 2-times expansion of the input n x n image by the inserting of a noisy pixel between each two pixels of original image. The noise is a uniform noise with a small dispersion and the same mean value that the original image or the result of the 2-times interpolation of the original n x n image.
2. Evaluation of the spectrum of expanded noisy $2n$ x $2n$ image using the iterative procedure based on the fact that the lowest n x n spectral coefficients should be taken on each iteration from the spectrum of original image.
3. Noise reduction from the $2n$ x $2n$ image corresponding to the obtained $2n$ x $2n$ spectrum.
4. High frequency correction on the image obtained on the step 3 with the goal to sharpen the smallest details smoothed during the noise reduction.

5. Final 2-times expansion of the input n x n image by the inserting of the corresponding pixels of image obtained on the step 4 between each two pixels of the original image.

The best way for the implementation of the steps 3 and 4 (noise reduction and high frequency correction) is multi-valued nonlinear filtering. As we convinced in the Section 5.3 it removes noise with maximal preservation of the signal. In the Section 5.4 it was shown how the same multi-valued filters may be effectively used for the frequency correction.

But first of all we have to obtain a more precise approximation of our resulting signal than (6.4.1). We will obtain it from the (6.4.1), taking into account that exact values of the signal $f(x, y)$ on the subdomain $\widetilde{A}$, and of its spectrum $F(u, v)$; u, $v \in \{0, 1, ..., n\text{-}1\}$, are known.

Let us build the following iterative and recursive process. Let $f_1(x, y) = g(x, y)$ be the starting approximation of the signal. Then

$$\widetilde{F}(u, v) = \Phi[f_1(x, y)], \qquad (6.4.2a)$$

and

$$F_1(u, v) = \begin{cases} F(u, v); \ u, v \in \{0, 1, ..., n\text{-}1\} \\ \widetilde{F}(u, v); \ u, v \in \{n, ..., 2n\text{-}1\} \end{cases} \qquad (6.4.2b)$$

will be the starting approximation of the spectrum. It is easy to obtain the following:

$$g_1(x, y) = \Phi^{-1}[F_1(u, v)]. \qquad (6.4.3)$$

We can obtain the next approximation $f_2(x, y)$ for our function from (6.4.3):

$$f_2(x, y) = \begin{cases} f(x, y), \ if \ (x, y) \in \widetilde{A} \\ g_1(x, y), \ if \ (x, y) \in A\backslash\widetilde{A}. \end{cases} \qquad (6.4.4)$$

Evidently, the nth approximation $f_n(x, y)$ for $f(x, y)$ may be obtained in the following way:

$$F_{n-1}(u, v) = \begin{cases} F(u, v); \ u, v \in \{0, 1, ..., n\text{-}1\} \\ \Phi[f_{n-1}(x, y)]; \ u, v \in \{n, ..., 2n\text{-}1\}, \end{cases} \qquad (6.4.5)$$

$$g_{n-1}(x, y) = \Phi^{-1}[F_{n-1}(u, v)], \qquad (6.4.6)$$

$$f_n(x, y) = \begin{cases} f(x, y), \ if \ (x, y) \in \widetilde{A} \\ g_{n-1}(x, y), \ if \ (x, y) \in A\backslash\widetilde{A}. \end{cases} \qquad (6.4.7)$$

So the equations (6.4.1) - (6.4.7) define the iterative process obtaining the best approximation for the signal $f(x, y)$ and its spectrum $F(u, v)$. It has been proven [Shaker & Steenart (1978)] that for continuous signals and for a similar iterative process

$$\int\limits_{-\infty}^{\infty}\int\limits_{-\infty}^{\infty}\left|F(u,\ v)-F_n(u,\ v)\right|^2 dudv < \int\limits_{-\infty}^{\infty}\int\limits_{-\infty}^{\infty}\left|F(u,\ v)-F_{n-1}(u,\ v)\right|^2 dudv,$$

and in such a case

$$\lim_{n\to\infty}\int\limits_{-\infty}^{\infty}\int\limits_{-\infty}^{\infty}\left|F(u,\ v)-F_n(u,\ v)\right|^2 dudv = 0.$$

It means that a process can't be infinite, because the standard deviation is decreased on the each step, and an iterative process is converged. It is evident that in the discrete case, which we consider here, the standard deviation also has to decrease, but it is impossible to obtain a precise proof of the fact that $\lim_{n\to\infty} f_n(x,\ y) = f(x,\ y)$, or that $\lim_{n\to\infty} F_n(u,\ v) = F(u,\ v)$ because the discrete functions are not analytical. Despite this fact experience has shown that the process, which is defined by the equations (6.4.1) - (6.4.7), is always stable. Thus, we will obtain the following:

$$f_n(x,\ y) = \begin{cases} f(x,\ y),\ if\ (x,\ y) \in \tilde{A} \\ f(x,\ y) + \tilde{s}(x,\ y),\ if\ (x,\ y) \in A\backslash\tilde{A} \end{cases}, \qquad (6.4.8)$$

where $\tilde{s}(x,\ y)$ is an additive noise, and

$$f_n(x,\ y) - f_{n-1}(x,\ y) = \begin{cases} 0,\ if\ (x,\ y) \in \tilde{A} \\ \varepsilon,\ if\ (x,\ y) \in A\backslash\tilde{A} \end{cases}, \qquad (6.4.9)$$

where ε is close to zero. So, we have to obtain the best approximation $\tilde{f}(x,\ y) = f_n(x,\ y)$ for $f(x,\ y)$, as a result of the iterative process (6.4.1) - (6.4.7). Experimental results show that (6.4.9) is true for any gray-scale images with uniform, or quasi-uniform distribution of the brightness with n (number of iterations) not greater than 7-8, and with use of Fourier, Cosine, or Walsh (ordered by Walsh) transforms as Φ in (6.4.2) - (6.4.3), and (6.4.5)-(6.4.6). Thus, the iterative process (6.4.1)-(6.4.7) is very short in practice.

According to the equation (6.4.8) $\tilde{f}(x,\ y) = f_n(x,\ y)$ contains an additive noise in the subdomain $A\backslash\tilde{A}$. This means that to complete a process of spectrum and signal restoration, we have to remove the noise, and to correct the high frequency after the noise removal. There are many different filters, which may be used for solution of noise removal problem. The multi-valued filters (MVF) described above in the Sections 5.3 and 5.4 are the best from our point of view because: 1) they remove noise with a maximal preservation of the useful signal (in, comparison for example, with order statistic filters); 2) it is possible to use the MVF not only for the noise removal, but for the high frequencies amplification, which is very important to save the smallest image details and boundaries; 3) The MVF may be implemented using CNN-MVN.

We will use the MVF (5.3.5) or more exactly, the template (5.3.7) for its implementation using CNN-MVN to remove the noise corrupting the image after procedure (6.4.1)-(6.4.7). Since the noise $\tilde{s}(x, y)$ (see (6.4.8)) corrupts only a part of the pixels, it is possible to use $w_{22} > 1$ in (5.3.7), which is better from the point of view of boundary preservation. Let $\tilde{\tilde{f}}(x, y)$ be the result of filtering the signal $\tilde{f}(x, y) = f_n(x, y)$ (see (6.4.8)). We can obtain a closes approximation for $f(x, y)$ in the following way:

$$\hat{f}(x, y) = \begin{cases} f(x, y), \ if \ (x, y) \in \tilde{A} \\ \tilde{\tilde{f}}(x, y), \ if \ (x, y) \in A \backslash \tilde{A} \end{cases}. \qquad (6.4.10)$$

Finally, we have to correct the high frequency part of the signal $\hat{f}(x, y)$, since noise removal has smoothened it. The multi-valued filter (5.4.10) and the corresponding template (5.4.12) will be used to make such a correction of the signal $\hat{f}(x, y)$, which has been obtained by (6.4.10). Let $\hat{\hat{f}}(x, y)$ be a result of this correction. To complete the process we have to do the following:

$$f^*(x, y) = \begin{cases} f(x, y), \ if \ (x, y) \in \tilde{A} \\ \hat{\hat{f}}(x, y), \ if \ (x, y) \in A \backslash \tilde{A} \end{cases}, \qquad (6.4.11)$$

where the signal $f^*(x, y)$ is the final approximation for our signal $f(x, y)$ of a size $2n \times 2n$ and defined on the domain A.

Evidently, it is possible to repeat the process defined by the (6.4.1) - (6.4.7), (6.4.10) - (6.4.11) and to obtain the approximation $f_2^*(x, y)$ for the signal of a size $4n \times 4n$ and so on. Each following image will have a double resolution in comparison with the previous one. This means that really the process defined by (6.4.1) - (6.4.7), and (6.4.10) - (6.4.11) leads to solution of the super resolution problem.

Let consider an example of application of the proposed method. The results of the double resolution of 256 x 256 testing image "Lena" using Fourier transform are presented in the Fig. 6.4.3. The iterative process (6.4.1) - (6.4.7) has converged after 5 iterations. A comparison of the resulting images obtained respectively, by this method (Fig. 6.4.3 (d)) and by bicubic interpolation (Fig. 6.4.3 (e)) shows the advantages of a new approach. The image from Fig. 6.4.3 (d) is much crisper than that in Fig. 6.4.3 (e). The standard deviation between the original 512 x 512 "Lena" and the resulting image from Fig. 6.4.3 (d) (this method) is equal to 3.17. The standard deviation between the original 512 x 512 "Lena" and the resulting image from Fig. 6.4.3 (e) is equal to 5.47 (bicubic interpolation). To illustrate the intermediate spectra and image approximations the corresponding results are presented in Fig. 6.4.4.

As it was mentioned above a same process may be repeated taking a previous resulting image as an input one, and an image with a two times higher resolution will be obtained in each following step. Fig. 6.4.5 contains 1024 x 1024 image "Lena" obtained by the same approach from the image presented in Fig. 6.4.3 (e).

The examples show the efficiency and advantages of the proposed approach to the super resolution problem.

To compare the presented approach with the bicubic interpolation the reader also can address to the Fig. 6.4.6. The graphic presentation of the amplitude of Fourier spectra corresponding to the 512 x 512 image "Lena" obtained by our approach (Fig. 6.4.3 d) 512 x 512 and image "Lena" obtained using bicubic interpolation are shown there. The high frequency part of spectra of the interpolated image is smoothed; a lot of the spectral coefficients are very close to zero. As a result the image obtained by bicubic interpolation is not sharp. At the same time the high frequency part of spectra corresponding to the image obtained by our approach is really restored with high precision. As a result the corresponding image is very sharp.

As it was mentioned above the second approach for the same problem is based on time series prediction using single MVN.

To simplify the considerations let consider a 1-D case. Let we have a signal f defined by its discrete values $f_1, f_2, ..., f_i, f_j, ..., f_N$ on the equal intervals. Our problem is to evaluate values

$$f_{i+s}, \quad i = 1, ..., N; \ i+s < j; \ s = 1/p, 2/p,..., (p-1)/p; \ p \in \{2, 4, 8, 16, ...\}.$$

For example, $p=4$ and we have to obtain $f_{i+0.25}, f_{i+0.5}, f_{i+0.75}$ between each f_i and f_j. A solution of the considered problem is reduced to the following steps: 1) Evaluation of the spectrum (Cosine or Walsh ordered by Walsh) of the signal f: $S_f = (s_1, ..., s_N)$; 2) Supposing that $s_i = g(s_{i-q}, ..., s_{i-1})$ (each spectral coefficient is a function of the q lowest ones, which are its predecessors), it is possible to train MVN by learning algorithm with rule (4.2.1) or (4.3.1) to implement the mapping that is described by the function g; 3) Extrapolation of the spectrum $\widetilde{S}_f = (s_1, ..., s_N, \widetilde{s}_{N+1}, ..., \widetilde{s}_{pN})$. Spectral coefficients corresponding to the highest frequencies are being predicted using MVN and the weighting vector obtained on the step 2; 4) Evaluation of the pN - dimensional inverse transformation and obtaining the unknown values of the signal f.

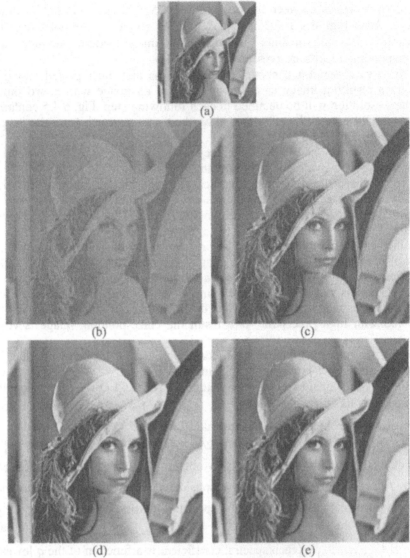

(a) - input 256x256 image; (b) - starting 512 x 512 approximation (signal + noise); (c) - result of the first iteration of the process (6.4.1) - (6.4.7); (d) - the final 512x512 image obtained by iterative process (5 iterations), noise reduction and high frequency correction using MVF and substitution (6.4.11) – standard deviation from the original 512 x 512 image is 3.17; (e) - the final 512x512 image obtained by bicubic interpolation - standard deviation from the original 512 x 512 image is 5.47.

Double resolution on 256 x 256 image "Lena" using iterative approximation of the noisy image spectra, and multi-valued filtering (approach 1)

Figure 6.4.3.

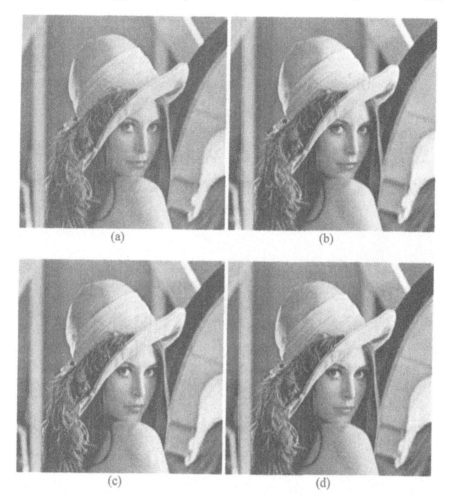

(a) - the noisy image (6.4.8), which is a final result of the iterative process (6.4.1) - (6.4.7) (4 iterations); (b) - the result of noise reduction from the image (a) using multi-valued filtering (filter (5.3.5), template (5.3.7), w_{22} =5.0); (c) - the image (6.4.10), which is a result of high frequency correction using multi-valued filtering (filter (5.4.10), template (5.4.12), G=16.0); (d) - the final result of the process obtained using substitution (6.4.11) the pixels of the original   256 x 256 image (see Fig. 6.4.3 (a))to the image (c) (the last image is the same that one in Fig. 6.4.3 (d)).

Figure 6.4.4. Intermediate and final results of the solution of the two times increase the resolution of 256 x 256 image "Lena"

4-times increase of the resolution of 256 x 256 image "Lena" using iterative
approximation of the noisy image spectra, and multi-valued filtering
(approach 1)
Figure 6.4.5.

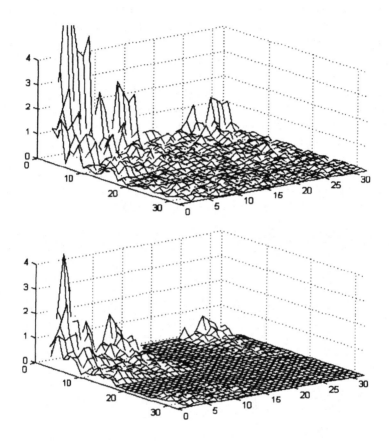

Amplitude of the Fourier spectra: image obtained by iterative process and
multi-valued filtering (top);
image obtained by bicubic interpolation (bottom).

Figure 6.4.6.

Thus, a main idea of the proposed solution is the assumption that higher frequency spectral coefficients may be expressed through the lower ones. In other words spectral coefficients ordered by frequency (from low to high) are considered here like time series (6.3.1), and are expressed one each other like (6.3.2). Since discrete spectra are not analytical functions it is impossible to obtain a precise proof of the fact that such a representation is true, or not. Actually, the rationale for the proposed solution is based on the following factors: 1) choice of the appropriate value of q on the second step (see above); 2) choice of the appropriate value of k (value of k-valued logic), and therefore, of interval for normalization of the spectral coefficients to create a learning set for MVN; 3) the time, which is needed for convergence of the learning algorithm. Evidently, speed of the learning is dependent on choice of q and k.

Let consider examples of Walsh ordered by Walsh spectra extrapolation presented in Fig. 6.4.7. We used MVN learning algorithm with Strategy 2 and learning rule (4.3.1) (see Sections 4.1, and 4.3).

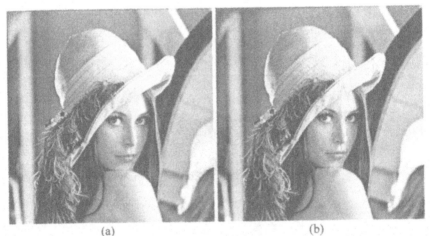

(a)   (b)

(a) - k=8192, q=64, standard deviation from original 512 x 512 image is 4.7;
(b) - k=4096, q=64, standard deviation from original 512 x 512 image is 5.1;

Figure 6.4.7. double resolution on 256 x 256 testing image "Lena" using prediction of the highest frequency coefficients of Walsh ordered by Walsh spectrum on MVN (approach 2)

A proposed approach has been checked for two values of k: 8192, and 4096. The lower values of k ensure quicker convergence of the learning algorithm. E.g., for k=4096 about 1 million iterations were necessary for convergence, but for k=8192 learning algorithm has converged after about 5 millions iterations; learning set in both cases consisted of 1024 elements corresponding to the highest frequency domain of the original 256 x 256 image. At the same time the higher values of k ensure much better precision for presentation of the spectral coefficients. As a result the quality of interpolated image is better. Unlucky chosen value of k involves "discretization artifacts" on the image form Fig. 6.4.7 (b).

A comparison of the proposed approaches to solution of super resolution problem shows that the first one is universal. The second one is good for more smooth spectra, when learning converges rather quickly and normalization (and further restoring) of the spectra values don't add artifacts to the restored spectrum (image).

It should be mentioned also that both approaches give good results, when the low frequency part of the spectra, which should be extrapolated, is not very small. For example, the quality of the increased resolution of the input images with sizes 256 x 256 and more is always high. For the smaller sizes of input images the quality of the results is little bit lower. It is reduces with a decrease of the input sizes of the images.

Use of MVN for both approaches is very important from our point of view. Multi-valued filtering, and its CNN-MVN implementation is a keyword point of the first approach. The high functionality of MVN and quickly converging learning algorithm for MVN are the key elements of the second approach.

All the considered results may be generalized for the color images. Such a generalization is natural: each color channel has to be processed separately using algorithms for gray-scale images just described above.

5.   CONCLUSIONS

Different MVN-based neural networks and their applications have been considered.

Associative memories based on CNN-MVN and neural network with MVN and random interconnections have been considered. High capacity of these networks has been estimated. It was shown that both models are more simple and efficient than Hopfield-like associative memory.

The single-layered MVN based neural network for image recognition using analysis of the lower part of the orthogonal spectra has been proposed. Its high efficiency has been shown on the example of face recognition.

The ability of MVN to predict time series has been shown.

Two solutions of the super resolution problem based on approximation of the spectra using multi-valued filtering, and prediction of the high frequency spectral coefficients on MVN have been considered.

Chapter 7

Conclusions, Open Problems, Further Work

We would like to conclude with a brief summary of the main results of the book, and to note prospectively some suggested directions for further work.

We started this book with the aim to show the new possibilities opened by neurons with complex-valued weights and activation functions dependent of the argument of the weighted sum. We very much hope that this book and the results presented above will stimulate further research work in the considered field.

The following results are the most important from our point of view.

1. Theory of multi-valued neurons and multiple-valued (k-valued) threshold functions.

- A coding of the values of k-valued logic using k^{th} roots of unity has been proposed.
- A notion of the k-valued threshold function over the complex numbers field has been introduced.
- The multi-valued neuron, which is a neural element implementing a mapping described by k-valued threshold function using a nonlinear activation function of argument of the weighted sum, has been proposed.
- k-separation of the n-dimensional complex space has been investigated.
- Method of MVN synthesis has been shown to reduce to the solution of linear programming problem.
- The features of the k-valued threshold functions over the complex numbers field have been considered.

- The quickly converging learning algorithm for MVN with two learning strategies and two linear learning rules has been proposed. Its convergence has been proven for both learning rules.

2. Theory of universal binary neurons and *P*-realizable Boolean functions.

- A notion of the *P*-realizable Boolean function over the complex numbers field has been introduced.
- The universal binary neuron, which is a neural element implementing a mapping described by *P*-realizable Boolean function using a nonlinear activation function of argument of the weighted sum, has been introduced.
- It was shown that a functionality of the UBN is always higher than a functionality of the threshold neurons.
- It was shown that the activation function P_B may be defined in such a way that the corresponding UBN will implement all the Boolean functions of the given number of variables.
- The features of the *P*-realizable Boolean functions have been considered. The synthesis of the UBN has been reduced to the solution of a system of linear algebraic equations.
- A notion of the *P*-realizable Boolean function has been generalized for the finite field, residue class ring, and arbitrary algebra *A*. Methods of synthesis of the UBN over finite algebras have been considered.
- It was shown how a notion of multiple-valued threshold function may be generalized by the same way.
- The quickly converging learning algorithm for UBN with the two linear learning rules has been proposed, and its convergence has been proven.

3. Applications of UBN and MVN

- It was shown that UBN and MVN may be successfully used as the basic neurons in cellular neural networks (respectively, CNN-UBN and CNN-MVN).
- A problem of the precise edge detection on the binary and gray-scale images has been considered. It was proposed to solve such a problem by description of edge-detection using Boolean function and its implementation using CNN-UBN. A problem of impulsive noise filtering has been solved in the same way.
- It was shown that detection of edges corresponding to upward, downward and disjunctive (upward, downward, or upward and downward together) jumps of brightness must be considered separately and should be described by the different Boolean functions.

- Edge detection by narrow directions has been considered. This problem also has been solved using description by the non-threshold Boolean functions and their implementation using CNN-UBN.
- It was shown that the gray-scale images might be processed by Boolean functions using separate processing of their binary planes.
- Multi-valued nonlinear filters based on non-linearity of the MVN activation function have been proposed. It was shown that they could be implemented using CNN-MVN.
- High efficiency of multi-valued filters for noise reduction and their advantages in comparison with order-statistic filters have been shown.
- Solution of the frequency correction problem, which leads to the extraction of image details using multi-valued filtering has been performed.
- Associative memories based on CNN-MVN and neural network with MVN and random interconnections have been considered. High capacity of these networks has been estimated. It was shown that both models are more simple and efficient than Hopfield-like associative memory.
- The single-layered MVN based neural network for image recognition using analysis of the lower part of the orthogonal spectra has been proposed. Its high efficiency has been shown on the example of face recognition.
- The ability of MVN to predict time series has been shown.
- Two solutions of the super resolution problem based on approximation of the spectra using multi-valued filtering, and prediction of the high frequency spectral coefficients on MVN have been considered.

4. Development of the mathematical background of neural networks.
- All features of multiple-valued threshold functions and *P*-realizable Boolean functions, methods of MVN and UBN synthesis have been investigated involving the group characters properties.
- The efficiency of the discrete spectral analysis methods for MVN and UBN synthesis (Reed-Muller, Walsh, Chrestenson transformations have been used) has been proven.

We hope that there are enough results to be able to speak about the new and powerful types of the neural elements, respectively, the multi-valued and universal binary neurons. They are similar to each other: both use the complex-valued weights, their activation functions are the functions of the argument of the weighted sum. Quickly converging learning algorithms based on similar learning rules have been introduced for both neurons.

Several types of neural networks based on UBN and MVN have been considered and their high efficiency for the solution of the different applied problems has been shown.

We hope to involve new researches to the work in the field developed in this book. We see several new directions for the further work.

What open problems are clearly visible? Let present some of them:

- Elaboration of the learning algorithms for UBN and MVN over the finite field and residue class ring.
- Investigation of the effective methods for solution of the UBN and MVN synthesis reduced to solution of the systems of linear algebraic equations.
- Seeking for necessary conditions for *P*-realization of all Boolean function of given number of variables.
- VLSI implementation of MVN and UBN.
- Investigation of the multi-layered MVN-based neural networks.
- Investigation of the new Boolean filters for image processing, further investigation of multi-valued filtering.
- Application of MVN-based neural networks to pattern recognition and related problems.

This book should be a good background for solution of all the mentioned problems. Evidently, more problems may be formulated, especially in relation to the applications of MVN, UBN and networks based on them. It is clear that we did not consider all the possible applications. Only the applications that are closer to our own interests have been considered. But e.g., at least in pattern recognition MVN-based neural networks may be applied to the solution of other problems than the face recognition considered in the book. It is also clear that the different applied problems related to implementation of the non-threshold Boolean functions may be reconsidered taking into account the ability of UBN to implement arbitrary (not only threshold) Boolean function on the single neuron.

Finally, we hope that our modest contribution to the general theory of neural networks and their applications will give some impulse to this field to move forward to the new developments in theory and practice.

REFERENCES

Ahmed N., Rao K.R. (1975) *Orthogonal Transforms for Digital Signal Processing*, Springer-Verlag, 1975.

Aizenberg I.N. (1984) About One Approach of the Coding of Binary Signals Over the Finite Fields, Kibernetika (Cybernetics), No 5, pp. 118-119 (in Russian).

Aizenberg I.N. (1985) Model of the Element With Complete Functionality. Izvestia Akademii Nauk SSSR, Technicheskaia Kibernetika (The News of the Academy of Sciences of the USSR, Technical Cybernetics), 1985, No 2, pp. 188-191 (in Russian).

Aizenberg I.N. (1991) The Universal Logical Element over the Field of the Complex Numbers, Kibernetika (Cybernetics), No 3, pp.116-121 (in Russian).

Aizenberg I.N. (1997a) Processing of Noisy and Small-Detailed Gray-Scale Images Using Cellular Neural Networks, Journal of Electronic Imaging, Vol. 6, No 3, pp. 272-285.

Aizenberg I.N. (1997b) Multi-Valued Non-Linear Filters and their Implementation on Cellular Neural Networks, Frontiers in Artificial Intelligence and Applications, Vol. 41."Advances in Intelligent Systems" (Ed. F.C.Morabito) IOS Press, Amsterdam-Berlin-Oxford-Tokyo-Washington DC, pp. 135-140.

Aizenberg I.N. (1998) Some Image Processing Algorithms Based on Neural Networks Technology, SPIE Proceedings, Vol. 3402, pp. 382-391.

Aizenberg I.N. (1999) Neural Networks Based on Multi-Valued and Universal Binary Neurons: Theory, Application to Image Processing and Recognition, Lecture Notes in Computer Science, Vol. 1625 "Computational Intelligence: Theory and applications" (Ed. - B. Reusch), Springer-Verlag, pp. 306-316.

Aizenberg I.N., Yaroslavsky L.P. (1989) Implementation of the Two-stage Method of Digital Picture Coding by One-Dimensional Cosine Transform, Proceedings of the International Conference on Image Processing and Pattern Recognition, London, IEE.

Aizenberg,I.N., Aizenberg N.N. (1997) "Universal Binary and Multi-Valued Neurons Paradigm: Conception, Learning, Applications", Lecture Notes in Computer Sciense, (Eds. - J.Mira, R.Moreno-Diaz, J.Cabestany - Eds.), Vol. 1240, Springer-Verlag, pp. 463-472.

Aizenberg I.N., Vandewalle J. (1997) Application of the Neural Networks from Multi-Valued Neurons for Solution of Pattern Recognition and Non-Linear Filtering Problems, Proceedings of International. Workshop "Digital metodologies and applications for multi-media and signal processing" (DMMS'97), Budapest, October, 27-28, 1997, Panem-Press, Budapest, pp. 27-36.

Aizenberg,I.N., Aizenberg N.N. (1998) Application of the Neural Networks Based on Multi-Valued Neurons in Image Processing and Recognition, SPIE Proceedings, Vol. 3307, 1998, pp. 88-97.

Aizenberg I.N., Vandewalle J. (1998) Multi-Valued Non-Linear Filters, Multi-Valued Neurons and Their Application to Extrapolation of the Orthogonal Spectra, Proceedings of the IEEE Benelux Signal Processing Symposium SPS-98, Leuven, Belgium, March 26-27, pp. 15-18.

Aizenberg I.N., Aizenberg N.N., Agaian S., Astola J.T., Egiazarian K. (1999) "Nonlinear Cellular Neural Filtering for Noise Reduction and Extraction of Image Details, SPIE Proceedings, Vol. 3646, pp.100-111.

Aizenberg I.N., Aizenberg N.N. (1999) Pattern Recognition Using Neural Networks Based on Multi-Valued Neurons, Lecture Notes in Computer Science, Vol. 1607 "Engineering Applications of Bio-inspired Artificial Neural Networks" (Eds. - J.Mira, J.V.Sanchez-Andrres), Springer-Verlag, pp. 383-392.

Aizenberg N.N., Ivaskiv Yu.L., Pospelov D.A. (1971a) About One Generalization of Threshold Function, Dokladi Akademii Nauk SSSR (The Reports of the Academy of Sciences of the USSR), Vol. 196, No 6, pp. 1287-1290 (in Russian).

Aizenberg N.N., Ivaskiv Yu.L., Pospelov D.A., Hudiakov G.F. (1971b) Multiple-Valued Threshold Functions. Boolean Complex-Threshold Functions and Their Generalization, Kibernetika (Cybernetics), No 4, pp. 44-51 (in Russian).

Aizenberg N.N., Ivaskiv Yu.L., Pospelov D.A., Hudiakov G.F. (1973) Multiple-Valued Threshold Functions. Synthesis of the Multiple-Valued Threshold Element, Kibernetika (Cybernetics), No 1, pp. 53-66 (in Russian).

Aizenberg N.N., and Ivaskiv Yu.L. (1977). *Multiple-valued threshold logic*, Naukova Dumka, Kiev (in Russian).

Aizenberg N.N. (1978) The Convolution Spectrum of Discrete Signals in an Arbitrary Basis, Dokladi Dokladi Akademii Nauk SSSR (The Reports of the Academy of Sciences of the USSR), Vol. 241, No 3, pp. 551-554 (in Russian).

Aizenberg N.N., Trofimluk O.T. (1981) Conjunctive Transformation of Discrete Signals and their Application for Construction of Tests and Recognition of Monotonicity of Boolean Functions, Kibernetika (Cybernetics), No 5, pp. 138-139 (in Russian).

Aizenberg N.N., Butakov V.D., Krenkel T.E., Harbash Ya.G. (1984) The Fresnel functions and transforms for linear non-degenerate transforms, Radiotekhnika i Elektronika (Radiotechnique and Electronics), vol. XXIX, No 4, pp. 698-704 (in Russian).

Aizenberg N.N. (1991) "Fast Methods of the Solution of Boolean Equations Systems and Equations Systems over the Real Numbers Field with Boolean Restrictions", Kibernetika ("Cybernetics"), No 3, pp. 126 - 127 (in Russian).

Aizenberg N.N., Aizenberg I.N. (1991) Model of the Neural Network Basic Elemets (Cells) with Universal Functionality and various of Harware Implementation, Proceedings of the 2-nd International Conference "Microelectronics for Neural Networks", Kyrill & Methody Verlag, Munich, 1991, pp. 77-83.

Aizenberg N.N., Aizenberg I.N. (1992). CNN Based on Multi-Valued Neuron As a Model of Associative Memory for Gray-Scale Images, Proceedings of the 2-d International Workshop on Cellular Neural Networks and their Applications (CNNA-92), Munich, October1992, pp. 36-41.

Aizenberg N.N., Aizenberg I.N., Pyshnyi M.Ph. (1993) Multi-valued and Universal Basic Elements for CNN: Mathematical Model and Hardware Implementation, Proceedings. of the 3-d International Conference "Microelectronics for Neural Networks", Edinburgh, UK, April, 1993, pp. 91-96.

Aizenberg N.N, Aizenberg I.N. (1993) Fast Converged Learning Algorithms for Multi-Level and Universal Binary Neurons and Solving of the Some Image Processing Problems, Lecture Notes in Computer Science, (Ed.-J.Mira, J.Cabestany, A.Prieto), Vol.686, Springer-Verlag, pp. 230-236.

Aizenberg N.N., Aizenberg I.N. (1994a) Neural Networks based on Universal and Multi-Valued Neurons and Their Application to Solving of the Some Problems of Image Processing and Pattern Recognition, Proceedings of the "COST 229" International Workshop on Adaptive Systems, Intelligent Approaches, Massively Parallel Computing and Emergent Techniques in Signal Processing and Communications", Bayona (Vigo), Spain, October, 1994, Publication of the University of Vigo and Politechnico de Madrid, pp. 223-228.

Aizenberg N.N., Aizenberg I.N. (1994b) CNN-like Networks Based on Multi-Valued and Universal Binary Neurons: Learning and Application to Image Processing, Proceedings of the Third IEEE International Workshop on Cellular Neural Networks and their Applications, Rome, Italy, December 18-21, 1994, pp. 153-158.

Aizenberg N.N,. Aizenberg I.N, Belikova T.P. (1994) Extraction and Localization of Important features on Gray-Scale Images: Implementation on the CNN. Proceedings of the Third IEEE International Workshop on Cellular Neural Networks and their Applications, Rome, Italy, December 18-21, 1994, pp. 207-212.

Aizenberg N.N., Aizenberg I.N., Krivosheev G.A. (1995a) Multi-Valued Neurons: Learning, Networks, Application to Image Recognition and Extrapolation of Temporal Series", Lecture Notes in Computer Science, (Eds. - J.Mira, F.Sandoval), Vol. 930, Springer-Verlag, pp.389-395.

Aizenberg N.N., Aizenberg I.N., Krivosheev G.A. (1995b) Neural Networks Based on Multi-Valued Neurons: Learning and Application to Image Processing and Recognition", Komputernaja Optika ("Computer Optics"), Vol. 14-15, part 1, pp. 179-186 (in Russian).

Aizenberg N.N., Aizenberg I.N., Krivosheev G.A. (1996a) CNN based on Universal Binary Neurons: Learning algorithm with Error-Correction and Application to Impulsive-Noise

Filtering on Gray-Scale Images, Proceedings of the Fourth International Workshop on Cellular Neural Networks and their Applications, Seville, Spain, June 24-26, 1996, pp. 309-314.

Aizenberg N.N., Aizenberg I.N., Krivosheev G.A. (1996b) Multi-Valued and Universal Binary Neurons : Mathematical Model, Learning, Networks , Application to Image Processing and Pattern Recognition, Proceedings of the 13-th International Conference on Pattern Recognition, Vienna, Austria, August 25-30, 1996, track D, IEEE Computer Society Press, pp.185-189.

Aizenberg I.N., Aizenberg N.N., Vandewalle J. (1998a) Precise Edge Detection: Representation by Boolean Functions, Implementation on the CNN. Submitted to the Proceedings of the Fifth IEEE International Workshop on Cellular Neural Networks and their Applications, London, UK, April 14-17, 1998, pp. 301-306.

Aizenberg I.N., Aizenberg N.N., Vandewalle J. (1998b) Solution of the Super Resolution Problem by Multi-valued Non-Linear Filtering, and its Implementation Using Cellular Neural Networks, Proceedings of the Fifth IEEE International Workshop on Cellular Neural Networks and their Applications, London, UK, April 14-17, 1998, pp. 353-358.

Aizenberg I.N., Aizenberg N.N., Vandewalle J., Gotko E. (1998c) Ultrasound Medical Image Processing Using Cellular Neural Networks. Proceedings of the 6th European Symposium on Artificial Neural Networks, Brugge, Belgium, April 22-24, 1998, D-facto publications, Brussels, 1998, pp. 315-320.

Aleksander I., and Morton H. (1990) *An Introduction to Neural Computing*. London: Chapman & Hall.

Argyle E. (1971) Techniques for Edge Detection. Proceedings of IEEE. Vol. 59, No 2, pp. 285-287.

Asar A., and McDonald J.R. (1994) A Specification of Neural Network Applications in the Load Forecasting Problem. IEEE Transactions on Control Systems Technology, Vol. 2, pp. 135-141.

Askerov Ch. I. (1965) Some Generalizations of Threshold Elements, Networks for Information Transmission an their Automation (Ed. V. Siforov), Nauka, Moscow, pp. 126-134 (in Russian).

Astola J., and Kuosmanen P. (1997) *Fundamentals of Nonlinear Digital Filtering*. CRC Press, Boca Raton, N.Y.

Banzhaf W. (1987) Towards Continuous Models of Memory, Proceedings of the 1-st IEEE International conference on Neural Networks, San-Diego, June 1987, Vol. 2, pp. 223-230.

Barron A.R. (1993) Universal Approximation Bounds for Superposition of a Sigmoidal Function, IEEE Transactions on Information Theory, Vol. 39, pp. 930-945.

Belikova T.P. (1990) Simulation of the Linear Filters in the Problems of Medical Diagnostics, in *Digital Image and Fields Processing in the Experimental Researches*, (I. Ovseevich - Ed), Nauka Publisher House, Moscow, pp. 130-152 (in Russian).

Bovik A.C., Huang T.S., Mudson D.C. (1983) A Generalization of Median Filtering Using Linear Combinations of Order Statistics', IEEE Transactions on Acoustic, Speech, Signal Processing , Vol.31, pp. 1342-1349.

Chua L.O., and Yang L. (1988) Cellular Neural Networks: Theory & Applications, IEEE Transactions on Circuits and Systems, Vol. 35, No 10, 1257-1290.

Chrestenson H.E. (1955) A Class of Generalized Walsh Functions, Pacific Journal of Mathematics, Vol. 5, pp. 17-31.

Chow C.K. (1961a) Boolean Functions Realizable with Single Threshold Devices. Proc. IRE, Vol. 39, pp. 370-371.

Chow C.K. (1961b) On the Characterization of Threshold Functions. Proc. Ann. Switching Circuit Theory and Logical Design, pp. 34-38.

CNNA (1990) Proceedings of the First IEEE International Workshop on CNN and their applications, (Ed. T.Roska, L.Chua) Budapest, Hungary, December 1990.

CNNA (1992) Proceedings of the Second IEEE International Workshop on CNN and their applications, (Ed. J.A.Nossek) Munich, Germany, October 1992.

CNNA (1994) Proceedings of the Third IEEE International Workshop on CNN and their applications, (Ed. V. Cimagalli) Rome, Italy, December 1994.

CNNA (1996) Proceedings of the Fourth IEEE International Workshop on CNN and their applications, (Ed. A.Rodriguez-Vaskez) Seville, Spain, June 1996.

CNNA (1998) Proceedings of the Fifth IEEE International Workshop on CNN and their applications, (Ed. V.Tavsanoglu) London, UK, April 1998.

Crounse K.R., Roska T., and Chua L.O. (1998) Practical Halftoning on the CNN Universal Machine, Proceedings of the Fifth IEEE International Workshop on Cellular Neural Networks and their Applications, London, UK, April 14-17, 1998, pp. 337-342.

Dantzig B.G. (1963) *Linear programming and extensions*. Princeton University Press, Princeton, New Jersey.

Dertouzos M.L (1965) *Threshold Logic*: A Synthesis Approach. The MIT Press, Cambridge, Mass., 1965.

Feller W. (1960) *An Introduction to Probability Theory and its Applications*, Wiley, New York .

Fleisher M. (1987) The Hopfield Model with Multi-Level Neurons, Proceedings of the AIP Conference on Neural Information Processing Systems, Denver, CO, 1987 (Ed. D.Anderson), pp. 278-279.

Fletcher R. (1987) *Practical Methods of Optimization*, second edition, Chichester and New York: John Wiley and Sons.

Foltyniewicz R. (1996) Automatic Face Recognition via Wavelets and Mathematical Morphology. Proceedings of the 13th International Conference on Pattern Recognition, Vienna, August 25-30, 1996, Track B, IEEE Computer Soc. Press, pp. 13-17.

Freeman W.J. (1975) *Mass Action in the Nervous System*. Academic Press, New York.

Georgiou G.M. and Koutsougeras C. (1992) Complex Domain Backpropagation, IEEE Transactions on Circuits and Systems CAS- II. Analog and Digital Signal Processing, Vol. 39, No 5, pp. 330-334

Good I.J. (1958) The Interaction Algorithm and Practical Fourier Analysis, J. Royal Stat. Soc. (London), Vol. B-20, pp. 361-372.

Goodman I.W. (1968) *Introduction to Fourier Optics*, McGraw-Hill, New York.

Grossberg S. (1976). Adaptive Pattern Classification and Universal Recoding. Part 1: Parallel Development and Coding of Neural Feature Detectors, Biological Cybernetics, Vol. 23, pp. 121-134.

Harrer H. and Nossek J.A. (1992) Discrete-Time Cellular Neural Networks, International Journal of Circuit Theory and Applications., Vol.20, pp. 453-467.

Hassoun M.H. (1995) *Fundamentals of Artificial Neural Networks*. MIT Press, 1995.

Haykin S. (1994) *Neural Networks. A Comprehensive Foundation*. Macmillan College Publishing Company, New York.

Hebb D.D. (1949) *The Organization of Behavior*, John Wiley & Sons, New York.

Hecht-Nielsen R. (1988) Neurocomputer Applications, in NATO ASI Series, Vol. F 41, Neural computers, (R.Eckmiller and Ch. V.d. Malsburg - Eds.), Springer-Verlag.

Hill F.J. and Peterson G.R. (1974) *Introduction to Switching Theory and Logical Design*, John Wiley and Sons, New York, 1974.

Hoffman K., Kunze R. (1971) *Linear Algebra*, second Ed.. Prentice-Hall, Englewood Cliffs, New Jersey.

Hopfield J. (1982) Neural Networks and Physical Systems with Emergent Collective Computational Abilities. Proceedings of the National Academy of Sciences of the USA. Vol. 79, pp. 2554 - 2558.

Hornik K, Stinchkombe M, and White H. (1989) Multilayer feedforward Networks are Universal Approximators", IEEE Transactions on Neural Networks, Vol. 2, pp. 359-366.

Huang T.S.(Ed.) (1975) *Picture Processing and Digital Filtering*, Springer-Verlag.

Jankowski S., Lozowski A., Zurada M. (1996) Complex-Valued Multistate Neural Associative Memory, IEEE Transactions on Neural Networks, Vol. 7, pp.1491-1496.

Justusson B.I. (1981) Median Filtering: Statistical Properties. Topics in applied Physics, Two-dimensional Digital Signal Processing II, (Ed. T.S.Huang). Vol. 43, Springer-Verlag, Berlin, pp. 161-196.

Kohonen T. (1977) *Associative memory - a Systemtheoretical Approach*. Springer-Verlag, Berlin.

Kohonen T. (1984) *Self-Organization and Associative memory*. Springer-Verlag, Berlin, 1984.

Kohring G.A. (1992) On the problems of neural networks with multistate neurons. Journal De Physique 1. Vol. 2, pp. 1549 - 1552.

Labunets V.G., Sitnikov O.P. (1975) Generalization of notion of k-valued threshold function over the Galois finite field, Izvestia Akademii Nauk SSSR, Technicheskaia Kibernetika (The News of the Academy of Sciences of the USSR, Technical Cybernetics), No 5, pp. 141-148 (in Russian).

Lang S. (1971) *Algebra*. Addison-Wesley, New York.

Lawrence S., Giles C. Lee, Ah Chung Tsoi and Back A.D. (1997) Face Rocognition: A Convolutional Neural-Network Approach, IEEE Transactions on Neural Networks. Vol. 8, pp. 98-113.

Lee C.-C. and Pineda de Gyvez J. (1996) Color Image Processing in a Cellular Neural Network Environment, IEEE Trans. On Neural Networks, Vol. 7, No 5, pp. 1086-1088.

Leung H., Haykin S. (1991) The Complex Backpropagation Algorithm", IEEE Transactions on Signal Processing, Vol.39, No 9, pp. 2101-2104.

McCulloch W.S. and Pits W. (1943) A Logical Calculus of the Ideas Immanent in Nervous Activity. Bull. Math. Biophys., 5, pp. 115-133.

Minsky M.L. (1961) Steps Towards Artificial Intelligence. Proceedings of the Institute of Radio Engineers, Vol. 49, pp. 8-30.

Minsky M.L. and Papert S.A. (1969) *Perceptron: An introduction to Computational Geometry*. MIT Press, Massachusets.

Muroga S. (1971) *Threshold Logic and Its Applications*. John Wiley & Sons, New York, 1971.

Moraga C. (1979) Extensions of Multiple-Valued Threshold Logic. Proceedings of 9th International Symposium on multiple-valued logic, IEEE CS Press, Silver Spring, Maryland.

Moraga C. (1989) Multiple-Valued Threshold Logic. Optical Computing Digital and Symbolic (Ed. - R.Arrathoon), Marcel Dekker, INC, New York and Basel.

Narendra K.S., Parthasarathy K. (1990) Identification and Control of Dynamical Systems Using Neural Networks, IEEE Transactions on Neural Networks, Vol.1, No 1, pp. 4-27.

Narendra K.S., Parthasarathy K. (1990) Gradient Methods for the Optimization of Dynamical Systems Containing Neural Networks, IEEE Transactions on Neural Networks, Vol.2, No 2, pp. 252-262.

Natarajan B.K. (1991) *Machine Learning: A Theoretical Approach*. Morgan Kaufmann, San Mateo, CA.

Nossek J.A. (1996) Design and learning with cellular neural networks, International Journal of Circuit Theory and Applications, Vol.24, pp. 15-24.

Novikoff A.B.J. (1963) On Convergence Proofs for Perceptrons, Stanford Research Institute. Report, prepared for the office of Naval Res. Under Contract No 3438 (00).

Peterson W.W. and Weldon E.J. (1972) *Error-Correcting Codes*. Second Edition, The MIT Press, Cambridge, Massachusetts and London.

Petkov N., Kruizinga P., Lourens T. (1993) Motivated Approach to Face Recognition, Lecture Notes in Computer Science. Vol. 686 (Eds. - J.Mira, F.Sandoval), Springer, pp.68-77.

Pitas I. and Venetsanopoulos A.N. (1990) Nonlinear digital filters: Principles and Applications. Kluwer Academic Publishers, Boston, 1990.

Pitas I. and Venetsanopoulos A.N. (1996) Nonlinear Mean Filters in Image Processing. IEEE Transactions on Acoustic, Speech, Signal Processing , Vol.34, pp. 573-584.

Poggio T., Girosi F. (1990) Networks for Approximation and Learning, Proceedings of the IEEE, Vol. 78, No.9, pp. 1481-1497.

Powell M.J.D. (1988) Radial Basis Function Approximations to Polynomials, Numerical Analysis 1987 Proceedings, pp. 223-241, Dundee, UK.

Pratt W.K. (1978) *Digital Image Processing*. John Wiley & Sons, N.Y.

Prewitt J.M.S. (1970) Object Enhancement and Extraction, in *Picture Processing and Psychopictorics* (B.Lipkin, A.Rosenfeld - Eds), Academic Press, New-York.

Rademacher H. (1922) Einige Sätze von Algemeinen Orthogonal Functionen, Math. Annalen, Vol. 87, pp. 122-138 (in German).

Ramaher U., Raab W., Anlauf J., Hachmann U., Beichter J., Bruls N., Weseling M., Sichender E., Manner R., Glass J., and Wurz A. (1993). Multiprocessor and Memory Architecture of the Neurocomputer Synapse-1. Proceedings of the 3-d International Conference on Microelectronics for Neural Networks. April 6-8, 1993, Edinburgh, UK, pp. 227-231.

Rekeczky C., Roska T. and Ushida A. (1996) CNN Based Self-Adjusting Nonlinear Filters Proc. of the Fourth IEEE International Workshop on Cellular Neural Networks and their Applications, Seville, Spain, June 1996, pp. 309-314.

Rosenblatt R. (1959) *Principles of Neurodynamics*, Spartan Books, New York.

Rosenfeld A. (1970) A Nonlinear Edge Detection Technique. Proceedings IEEE Letters. Vol. 58, No 5, pp. 814-816.

Rosenfeld A., Thurston M. (1971) Edge and Curve Detection for Visual Scene Analysis. IEEE Transactions on Computers. Vol. 20, No 5, pp. 562-569.

Roska T. and Chua L.O. (1993) The CNN Universal Machine: An Analogic Array Computer, IEEE Transactions on Circuits and Sysems. - II , Vol. 40, No 3, pp. 163-173.

Roska T. and Vandewalle J.(ed.) (1993) Cellular Neural Networks, John Wiley & Sons, New-York.

Rumelhart D.E., Hinton G.E., Williams R.J. (1986) Learning Representations By Back-Propagating Errors, Nature, Vol. 323, pp. 533-536.

Shaker Sabri M., and Steenart W. (1978) "An Approach to Band-Limited Signal extrapolation: the Extrapolation Matrix", IEEE Transactions on Circuits and Systems. Vol. 25, pp. 74-78.

Shi B.E. (1994) Order Statistic Filtering with Cellular Neural Networks", Proceedings of the Third IEEE International Workshop on Cellular Neural Networks and their Applications, Rome, December 1994, pp.441-444.

Si J., and Michel A.N. (1995) Analysis and Synthesis of a Class of Discrete-Time Neural Networks with Multilevel Threshold Neurons, IEEE Transactions on Neural Networks, Vol. 6, No 1, pp. 105-116.

Stark H. (1982) *Applications of optical Fourier transforms*, Academic Press, New York.

Sun X.Z., and Venetsanopoulos A.N. (1988)"Adaptive Schemes for Noise Filtering and Edge Detection by use of Local Statistics", IEEE Transactions on Circuits and Systems, vol. 35, pp. 57-69.

Suykens J.A.K., Vandewalle J.P.L., and De Moor B.L.R. (1996) *Artificial Neural Networks for Modeling and Control of Non-Linear Systems*. Kluwer Academic Publishers, Boston/Dordrecht/London.

Suykens J.A.K. and Vandewalle J.P.L. (eds.) (1998) *Nonlinear Modeling Advanced Balack Box Techniques* . Kluwer Academic Publishers, Boston/Dordrecht.

Suykens J., Lemmerling Ph., Favoreel W., De Moor B., Crepel M., and Briol P. (1996) Modelling the Belgian Gas Consumption Using Neural Networks. Neural Processing Letters. Vol. 4, pp. 157-166.

Schwartz L. (1967a) *Analyse Mathématicue, Cours Professe a L'ecole Polytechnique I.*, Hermann, Paris.

Schwartz L. (1967b) *Analyse Mathématicue, Cours Professe a L'ecole Polytechnique II.*, Hermann, Paris.

Tan S., Hao J., Vandewalle J. (1990) Cellular Neural Networks as a Model of Associative Memories, Proceedings of the 1990 IEEE International Workshop on CNN and their applications (CNNA-90), Budapest, December, 1990, pp. 23-26.

Tanaka M., Ikegami M., Imaizumi M., Shingu T., and Inoue H. (1996) Templates Design for High Quality Digital Images by Discrete Time Cellular Neural Network, Proceedings of the Fourth International Workshop on Cellular Neural Networks and their Applications, Seville, Spain, June 24-26, 1996, pp. 333-338.

Toth G., Foldesy P., and Roska T. (1996) Distance Preserving 1D Turing-Wave Models via CNN, Implementation of Complex-Valued CNN and Solving a Simple Inverse Pattern Problem (detection), Proceedings. of the Fourth IEEE International Workshop on Cellular Neural Networks and their Applications (CNNA-96), Seville, Spain, June 1996, pp. 109-114.

Turk M. and Petland A. (1991) Eigenfaces for Recognition. Journal of Cognitive Neuroscience. Vol. 3.

van der Smagt P.P. (1994) Minimization Methods for Training Feedforward Neural Networks, Neural Networks, Vol. 7, No 1, pp. 75-93.

van der Warden B.L. (1971) *Algebra I. Auflage der Modernen Algebra*, Springer-Verlag, Berlin/Heidelberg/New York.

Vilarino D.L., Cabello D., Balsi M., and Brea V.M. (1998) Image Segmentation Based on Active Contours using Descrite-Time Cellular Neural Networks, Proceedings of the Fifth IEEE International Workshop on Cellular Neural Networks and their Applications, London, UK, April 14-17, 1998, pp. 331-336.

Walsh J.L. (1923) A Closed Set of Orthogonal Functions, American Journal of Mathematics, Vol. 45, pp. 5-24.

Winder R.O. (1969) The Status of Threshold Logic. RCA Review, Vol. 30, pp. 235-318.

Zarandy A,. Stoffels A., Roska T, and Chua L.O. (1996) Morphological Operations on the CNN Universal Machine, Proceedings of the Fourth International Workshop on Cellular Neural Networks and their Applications, Seville, Spain, June 24-26, 1996, pp. 151-156.

Zurada J.M. (1992) *Introduction to Artificial Neural Systems*, West Publishing Company.

Zurada J.M., Cloete I., van der Poel E. (1994) Neural Associative Memories with Multiple Stable States. Proceedings of 3rd International Conference on Fuzzy Logic, Neural Nets, and Soft Computing Iizuka, Fukuoka, Japan, pp. 44-51.

Index

273